NOT
Your Mother's
FOOD
STORAGE

NOT
Your Mother's
FOOD
STORAGE

Store the Food You Use Every Day

KATHY BRAY *and* JAN BARKER

DESERET
BOOK

SALT LAKE CITY, UTAH

To all those families who want food storage

but don't know where to start

Special thanks to Julie Ann Pullman, Jan's daughter,

for her invaluable input and for asking the right questions

Library of Congress Cataloging-in-Publication Data

Bray, Kathy.
   Not your mother's food storage / Kathy Bray and Jan Barker.
     p. cm.
   Includes index.
   ISBN 978-1-60641-666-2 (paperbound)
   1. Food—Storage. I. Barker, Jan. II. Title.
   TX601.B733 2010
   641.4'8—dc22   2010011137

Printed in the United States of America
Malloy Lithographing Incorporated, Ann Arbor, MI

10 9 8 7 6 5 4 3

CONTENTS

WHY THIS PLAN?

This is *not* your mother's book on food storage. In it you won't find a "store it for some future time when you might need it in case of a disaster" program. Instead, you'll discover a program that will simplify your life *today*. Using this method to plan your food storage means you'll no longer be dependent on the supermarket: There will be no more quick runs to the grocery store to pick up something for dinner. (That alone will save you time and money.) Instead, you'll always have several meals planned, and you'll be able to use the food in your pantry to prepare them—anytime. These will be meals your family loves to eat. After all, why waste money storing food you may not even like?

For ten years, Jan has used this system to plan meals, buy food at sale prices, use it in her everyday cooking, and replace it when the items go on sale again. And, thankfully, it bears little resemblance to the once-hailed programs that led to buckets of stored wheat, honey, powdered milk, and oil collecting dust in the basement for years at a time.

When I was younger, those buckets—along with a generous supply of rice, beans, and freeze-dried goods—were our family's food storage plan. And, believe me, I never thought we'd actually have to eat it.

When we moved from northern California to Utah in 1986, we

filled nearly a quarter of the U-Haul truck with our accumulated food storage. The recession of the early 1980s had wiped us out financially, and we were starting over. With my husband's failing health, I became the main breadwinner.

After a month of looking, I was humbled and happy to get a job at a local print shop making five dollars an hour, even though I was used to earning thirty dollars per billed hour in California. But, we had our food storage, and that's what fed us for six weeks, during which time I bought only eggs and a few fresh vegetables. I don't know what we would have done were it not for our food storage. But, still, I struggled to make meals out of what we had available. And I definitely didn't look forward to eating those meals—with the exception of my homemade bread.

We were obedient and felt the Lord's hand over our lives during that time. We didn't starve. We ate what we had stored. And while we were grateful to have any food at all, the drastic change in our diet was hard to manage. As soon as we were able, what was left of our food storage was replaced by foods we actually liked to eat.

Jan learned this lesson, too, but in a different way. She and her husband, Jeff, started their food storage with a rather expensive system containing all of the staples plus cases of freeze-dried foods. One day, Jeff decided that if they were going to store all this food, they might as well figure out how to use it. They opened can after can, offering the contents to their somewhat bewildered children. No one wanted to eat any of it. Instead of raves, it brought shouts of "Yuck!" from everyone. There wasn't anything wrong with the food. Sure, the kids would have eaten it if starvation threatened. Maybe. But it was definitely not the taste they were used to. And if they weren't starving, they weren't going to eat it.

Now, we're not against freeze-dried food. It's nourishing and can be used in a lot of ways. But if you're going to store it, plan on getting your family used to eating it.

After that experience, Jan said, "Forget that! The kids would rather eat SpaghettiOs, so I'll store SpaghettiOs," which she purchased at the next case-lot sale. From that time on they would store only the food their family likes and is used to eating. She wrote down a menu for each day of the week, which became her guide for buying food storage. Now she always has food on hand to make the meals her family likes.

And Jan didn't stop at food. She decided she could use the same method with nonfood items. For example, she looked at how much toilet paper the family went through in a week and easily figured out what they needed to store for a month and then for a year.

Acquiring a three-month supply this way is easy. And you can always extend your supply to six months or even a year if you are financially able to. This book, however, is dedicated to helping you acquire and maintain a three-month supply of food and nonfood items.

When Jan explained this concept to me, I caught the vision of how easy it is to meld meal planning and food storage into a reliable, can-do system. Not to mention how much everyday stress it relieves to already have a meal plan in place. We wrote our ideas down on paper, and soon *Not Your Mother's Food Storage* was the result.

Because the system is based on foods your family likes to eat, you'll find that it works well both for those who like to cook from scratch and those who prefer to rely heavily on convenience foods. *Not Your Mother's Food Storage* is unlike any other food storage book because we help you plan your meals before you start buying the food. If you already have a lot of food storage, you can use the worksheets in the book to figure out how many meals you can make with what you already have stored. This is good for inventory, and you may find out that you're missing items from your food storage that are required to make some of your family's favorite meals.

Many of you probably already plan menus for a week and then create a shopping list based on those daily menus. This food storage plan works in much the same way: You plan a set number of breakfast, lunch, and dinner meals that you can rotate over a period of three months. The next step is to create a shopping list of storable food items that can be used to make the meals you've planned. Next, you calculate how much of each storable item you'll need to last for three months. Finally, you begin purchasing and *using* the food on your list, replacing it as you go, so you always have a rotating three-month supply of food and a happy, well-fed family.

Now is the time to change your attitude toward food storage. Start believing that the food you store is for eating now, not just for use during emergencies. Once your attitude has changed and your plan is in place, your family won't be able to tell whether they are eating from the food storage or not.

YOU CAN AFFORD TO DO IT

We understand that starting, building, and maintaining a three-month supply of food can seem overwhelming, not to mention expensive. But we've learned that a little time spent planning can go a long way. And that planning is exactly what makes it possible to afford to store enough food for three months and maintain that level of preparedness. We've also learned that once you make the decision to be obedient, the Lord always opens doors, providing a way for you to accomplish the task.

Whether you are just beginning your food storage program or have been storing food for years, *Not Your Mother's Food Storage* worksheets can be the answer to really knowing if you have enough of the right foods to make the meals your family will want to eat. You'll learn how to use the worksheets in the chapters that follow. But first, keep in mind these five food storage spending fundamentals:

1. Start Small. You can begin to build your food storage by simply purchasing an extra item or two when you see it on sale. You don't need to go out and purchase three months' worth of food in a single shopping trip. Even if all you do is spend ten dollars a month toward food storage, that's something. When supermarkets run case-lot sales, ten dollars can often buy a whole case of food.

2. Stick to a List. The key is to always work from your shopping list, which you will develop from your meal plans. This will help you make more efficient and systematic purchases. Whether you peruse the newspaper grocery ads or go online to check for specials and coupons, always keep your shopping list handy. When you find an item on sale that is on your list, plan to purchase a little more than you will immediately use. We are counseled not to go into debt for food storage, so just buy what you can afford, but be consistent about it. Your first goal may be to get enough stored for a one-month supply of meals and other essentials. You'll be surprised at how your food storage will grow when you work from a plan and consistently add to it.

3. Shop the Sales. When your local stores feature case-lot sales, use your shopping list to plan what you will buy. If you can't afford an entire case, buy what you can afford, or split a case with another family member or friend. Pay attention to the ads. Many times, the loss-leader items offered on the front page of grocery store ads are even cheaper than items you can buy by the case at case-lot sales. These are items that are normally priced at or below cost to bring customers into the store (where they hope you'll buy more than just what's advertised on the front page).

Just before major holidays, such as Memorial Day, Independence Day, or Labor Day, you'll find hot dogs and hamburgers at greatly reduced prices. That's the time to stock your freezer. You can often buy hamburger in family or value packs at a good price. Before freezing it, divide it into one-pound (or more) lots. During the Thanksgiving and Christmas holidays, baking goods are often on sale. Use those months to stock up on sugar, flour, baking powder, and so on.

Later on, we'll explain how to check your shopping list against the holiday sales to see where you can pay the lowest price for any items on your list. And we'll tell you why, as you put away your food items, to fill out your shopping list with the date and quantity of your purchases.

4. Consider Buying in Bulk. Compare bulk prices online to prices at your local retailer. A great Web site to check for bulk staples, such as oatmeal, salt, sugar, and other similar products, is Walton Feed (http://www.waltonfeed.com). Another site, Augason Farms, offers every type of food imaginable for food storage in varying sizes. Find their catalog at http://www.augusonfarms.com. You may also be able to find your own local source with even better prices. Here in the Mountain West, we like Alpine Food Storage (http://www.alpinefoodstorage.com).

Prices fluctuate on any food, but buying in bulk is usually considerably less expensive. For example, a twenty-five-pound bag of rolled oats will typically cost you in the range of ten to fifteen dollars, and it will last you a year or more. If you want, you can split a bag with someone else to bring down the initial cost. For that same price you would get only four or five canisters at the store.

5. Keep Cutting Coupons. We also encourage you to look into using coupons to save money on all of your grocery and nonfood shopping. Coupon clipping can be a fun family activity, and saving money will allow you to build your food storage at an even faster rate. One system we recommend is Savvy Shopper. Many states have similar systems, and a simple search on Google with your state's name and the words *grocery, coupons,* and *system* will help you find a site that suits your needs.

The Savvy Shopper system uses coupons in a way that can save you thousands—yes, we said thousands—of dollars on food and nonfood products every year. Go to http://www.savvyshopperdeals.com to sign up for a number of deals and coupon opportunities. At the Web site, click on your location to find out if this system is available where you live. You will get information on all of the sales at your local supermarkets and retailers. Sales are rated as to how good the deal is. And many times, you'll find links to coupons that will give you even more money off the sale price. You can typically save 50 to 80 percent over retail prices, and it is not uncommon for the store to wind up paying you money to buy an item when using this system! It's a free service, and we know many people who use it.

Having a system for using and efficiently organizing all those coupons is the advantage of a program like Savvy Shopper. Without a system, it's easy to cut out coupons and then forget to use them.

START NOW

You have many choices regarding how you will purchase your food storage: in bulk, during sales, using coupons, or all of the above. Start with a three-month plan, and then simply repeat the process to extend your food storage. Once you start, it will become a way of life, a valuable method of provident living.

We simply don't know what tomorrow may bring. One day you may be feeling financially secure, and the next day you may be out of a job or sideswiped by a major illness. Of course, natural disasters can also hit, but more often than not, it is a downturn in a family's economic situation—which can be brought on by any number of circumstances—that requires a family to rely on food storage for a time.

Even if you have already stored a lot of food, ask yourself if you have been using it so that your family will want to eat it when it's needed. You can use the *Not Your Mother's Food Storage* worksheets to plan meals that your family enjoys. Then, take stock of what you already have on hand. Follow the steps outlined in this workbook to find out what items you need to add to your food storage. Then you can begin using the food you have already stored instead of letting it get so old that you finally toss it.

We'll teach you how to adapt many favorite recipes by substituting items that can be stored. And we don't stop with food. The shopping lists you'll develop as you go through this workbook will also guide your nonfood purchases. Panic buying is often wasteful, and you could wind up with too much of one food and not enough of another to make the meals you have planned. By shopping with a plan, your money will be spent more efficiently and, in most cases, you won't have to pay retail prices.

It's a simple process, and we'll walk you through it, step by step. Now it's time to get started with your meal plans. We'll teach you how to plan breakfast meals first, then lunch and dinner. We'll show you our sample meal plans first, and then we'll provide a worksheet for you to practice planning your own meals.

You can also download free, full-sized meal planning and shopping list worksheets for breakfast, lunch, dinner, desserts and treats, staples and condiments, and nonfood items on our Web site: http://www.notyour mothersfoodstorage.com.

Chapter 2

PLANNING YOUR BREAKFAST MEALS

The trick to planning meals from food storage is to choose recipes that are simple to prepare and that don't use a lot of ingredients. Ask yourself, What does my family like to eat for breakfast that will store well? You can make a list of just a few meals to repeat throughout the week, or you can choose to prepare a different breakfast each day of the week. It's up to you. Following are some tips to help you start making a list of breakfast meal ideas.

- Cold cereal takes a lot of room. But if you have the space, go ahead and make the purchase. Watch the use-by dates closely and be sure to use and replace.

- Hot cereal stores well, is less expensive than cold cereal, and is more nutritious. For our sample breakfast meal plan, we chose oatmeal and Cream of Wheat (also known as *germade* or *farina* when purchased in bulk) for our cereal breakfasts.

- Consider pancakes and waffles, because you can buy mixes that require nothing but added water.

- If you love eggs, don't exclude them from your plan. We love eggs, and because eggs will keep in the refrigerator up to six weeks, we've also included an egg breakfast in our sample meal plan. If you continually use and replace eggs, you'll almost always have eggs on hand to use for meals and baking. Powdered eggs are an acceptable alternative to fresh eggs and, when prepared correctly, are delicious. In fact, many restaurants and cruise lines use powdered eggs for the scrambled eggs in their breakfast buffets.

- When planning any of your meals, be sure that the ingredients will store well. (For tips on storing fresh foods, such as whole potatoes and onions, see chapter 9, How to Store It Now That You Have It.)

THE FORMULA AND MEAL PLAN

Once you've made a list of meal ideas, the next step is to determine how many servings of each meal you will need to prepare. To do that, just follow this simple formula:

1. Decide how many times you want to have a particular meal during a one-month period.

2. Multiply that number by three.

3. Multiply that number by the number of people in your family. The result is the total number of servings needed for that particular meal for three months.

It's important to know the total number of servings needed when it comes to creating your shopping list, which we'll show you how to do later.

Here is a worksheet filled out with our sample plan for breakfast meals. You can download a blank worksheet from our Web site to make your own breakfast meal plan.

Breakfasts	# times per month	× 3 months	× number in family	= total number of servings
Oatmeal	8	24	4	96 servings
Cream of Wheat	6	18	4	72 servings
Pancakes and syrup	4	12	4	48 servings
Waffles and syrup	4	12	4	48 servings
Biscuits and gravy	4	12	4	48 servings
German pancakes	4	12	4	48 servings
Total meals	30			

Our meal plan worksheets make it easy to use our formula. Look at our sample breakfast meal plan and you'll see that we have planned six different breakfasts: oatmeal, Cream of Wheat, pancakes, waffles, biscuits and gravy, and German pancakes (an egg dish). Let's begin with oatmeal. We love oatmeal cooked with raisins, cinnamon, and a dash of vanilla, so we decided to have that meal twice each week, or eight times a month. So, we pencil in an 8 in the first column. Next, we multiply 8 by 3 months, which is the number of times we will serve oatmeal during the three-month period. We write the answer (24) in the second column. Then we multiply that number by how many people there are in the family. In our case, it's four. So, twenty-four servings multiplied by the number of people in our family equals ninety-six total servings. We write 96 in the last column. This formula will work for just about any food you want to store. Simple, right?

Note: Be sure that the total number of meals you'll serve each month equals thirty or thirty-one meals (the number of days in a month).

CONSIDER ADDING EXTRAS

You'll notice that our sample breakfast meal plan includes only a main dish. If your family is used to drinking juice with breakfast, consider supplementing your breakfast food storage with juice. Any kind of fruit or vegetable juice can add nourishment to a meal. Bottled juices store well unopened. For example, a bottle of V8 Vegetable Juice I recently bought has a shelf life of three years. If you have room in your freezer, you could also store frozen orange juice or apple juice concentrate. So, if you have the space and the funds, consider adding juices to your food storage plan.

WRITE DOWN THE PLAN

Now it's time to practice planning your breakfast meals. We suggest starting with breakfast meals first because they are the least expensive, and often the most important, meal of the day. Additionally, breakfast foods are often among children's favorite foods, so starting here is a good way to involve your kids in planning menus. Jan has found that some breakfast items, such as oatmeal and pancake mix, are so reasonably priced that she can easily buy more than a year's supply at once. This has the added benefit of enabling her to serve breakfast more than once a day if the family is really in a pinch.

Let's get started. Write down five to seven breakfast meals in the breakfast meal planning worksheet you downloaded from our Web site (http://www.notyourmothersfoodstorage.com).

This will get you started thinking in terms of meals instead of just "food storage items." Be sure that meals can be made from foods you store on a shelf, in your basement, in the freezer, or even in the refrigerator, such as eggs, which keep well for up to six weeks if refrigerated properly. If you can't think of a different breakfast, repeat a favorite breakfast as needed so that you have a breakfast planned for seven days, then decide how many times you will serve each breakfast during one month. From there just follow the formula on the worksheet.

CREATE A BREAKFAST SHOPPING LIST

Now that you have planned your breakfast meals, you're ready to create a shopping list, which will be your guide as you shop for food storage. It will tell you how much you need to store of each item required to prepare the meals in your breakfast meal plan. So, what would you purchase for the breakfast meals you've planned? If you've used the formula, you should know how many servings of each meal you need to store for a three-month supply. We've said that this is a "use and replace" plan; by that we mean use it and replace it *before* you run out. When you see you're running low on an item, watch for a sale and buy extra. If you buy extra when an item is on sale, you'll never have to pay full price. However, if you run out of an item you use all of the time, you'll be forced to pay full retail. Who wants to do that? The idea is to build *and maintain* a three-month supply of food and other essential items.

To help you create your shopping list, we'll walk through the process we use by reviewing the sample breakfast plan meal by meal. We'll start with oatmeal.

Oatmeal: One-half cup of dry oatmeal cooks up to 1 cup of cooked oatmeal, which equals one serving. According to the sample breakfast meal plan, we need ninety-six servings of oatmeal to last our family of four three months. Each serving is ½ cup, so the math tells us we need 48 cups of dry oatmeal. (See how important it is to know the total number of servings that you need for each planned meal?)

You can purchase oatmeal from a local retailer or buy it in bulk. Check the prices and the label before buying. The label will tell you how many servings come in a package and how big the serving size is. This information may surprise you. Sometimes, to make a food appear lower in calories, manufacturers make the serving size quite small. Make sure your math accounts for the serving size your family considers reasonable.

With any item, consider whether or not you might use the ingredient for something other than breakfast. For example, I like to have extra oatmeal on hand for cookies and a number of other favorite recipes. I like to put oatmeal in my meatloaf as an extender (rather than bread or bread crumbs) and in my chocolate chip cookies for extra nourishment. If you do find that you'll want the ingredient for uses other than breakfast, add another package to the list.

If you buy oatmeal at the grocery store, a 42-ounce (2 lb. 10 oz.) canister of old-fashioned or quick-cooking oats will give you thirty servings. If you find the individual canisters of oats at a good price, you will need to purchase at least four of them to have enough to serve oatmeal for breakfast twice each week for three months.

Jan and I buy oatmeal in bulk. I store it in buckets, filling a store-bought oatmeal canister for my pantry as needed. Jan dry packs oatmeal in #10 cans and brings in one can at a time to put in her pantry. Either way, it gives us a usable amount in the pantry for everyday use.

Before you move on to listing items for the next meal, consider any other ingredients you might need to prepare this meal. For example, we like to eat our oatmeal with a little vanilla, a handful of raisins, and a sprinkling of cinnamon. Cinnamon and vanilla are staples for any food storage program, so you don't really need to figure out how much to store for each meal. You'll always use them for more than just breakfast, so store as much as you can afford. (Staples are discussed in chapter 6, Storing Staples and Condiments.) Raisins, however, might be a different story. Maybe the only time you do eat them is with oatmeal. If that's the case, do some quick math to figure out how much you'll need to buy to get ninety-six servings.

Other hot cereals: For other kinds of hot cereals, such as Cream of Wheat or Malt-o-Meal, determine the amount you will need to buy using the same formula we used for the oatmeal. Jan buys germade (Cream of Wheat) in bulk and keeps a usable quantity handy in her pantry. It takes 3 tablespoons of germade to make one serving, so you can see that a little of that type of cereal will go a long way. If you buy it by the box as Cream of Wheat, just check the number of servings per box and use the same method we did for oatmeal to know how many boxes to store. Write the total on your shopping list. And don't forget to plan for sugar or honey and milk to go with cereal.

Pancakes and waffles: If pancakes and/or waffles are part of your planned meals, choose a mix such as Krusteaz, which just mixes with water. Ten-pound bags can be purchased at wholesale shopping clubs such as Costco or Sam's Club. Remember to look at how many servings the package contains and make sure that the serving size fits your family's appetite. In our sample plan, we would need enough pancake and waffle mix for ninety-six servings of waffles and pancakes. If your kids think a serving is three pancakes, but the package says a serving is one pancake, you'll need to do the math to make the adjustments necessary to meet your family's needs.

Don't forget butter and syrup. Jan buys 10 pounds of butter at a time, or more when it's on sale, and freezes it. For syrup, you can buy the gallon size at a good price at most wholesale shopping clubs. We typically use 1 cup of syrup for each breakfast of pancakes or waffles. There are 16 cups in a gallon, so we know that 1 gallon of syrup will last through 16 breakfast meals, or 64 servings (16 x 4 family members). According to our breakfast meal plan, we would need to have enough syrup for ninety-six servings, so we know that if we store 2 gallons of syrup, we'll have enough and to spare for three months. A less expensive solution to buying and storing syrup would be to make your own. All it takes is water, sugar, and maple flavoring, and because you will already be storing sugar, it takes less storage space and is always hot when served. Recipes for flavored syrups are included in chapter 11, Recipes.

Biscuits and gravy: If you decide to include biscuits and gravy as one of your breakfasts, like we did, you can choose to store gravy packets, or make your own gravy from the sausage drippings. Gravy packets, which make 2 cups of gravy, regularly go on sale for a dollar or less. Two cups is plenty for our family of four. So, we'll need one packet for every four servings. We need forty-eight servings, so we'll need twelve packets. Of course, we can make gravy from scratch using the recipe for country gravy in our recipe section.

For the biscuits, you'll just need flour, shortening, salt, baking powder and milk, or you can store a baking mix such as Bisquick, which keeps well for at least three months. Check the use-by date on the package, and store at a cool temperature for maximum freshness. Another alternative is to buy biscuit mix in bulk, which has a longer storage life.

If you choose to buy a baking mix, add it to your shopping list. Ingredients for making your own baking mix should go on your staples/condiments shopping list, because you'll use them for many other meals.

German pancakes: Our sample breakfast plan calls for German pancakes once each week. It's a special breakfast that our family loves, and it can be served several ways. It takes 6 eggs, 1 cup flour, 1 cup milk, and salt. (The recipe for German Pancakes is in chapter 11.) Jan keeps 4 to 6 dozen eggs on hand at any given time, because her family goes through a dozen or more each week. I also make sure I have plenty of eggs on hand. Milk and flour are staples that you will always have on hand. See tips on storing eggs in chapter 9, How to Store It Now That You Have It.

The breakfast shopping list, based on our sample plan, would look something like this:

BREAKFAST MASTER SHOPPING LIST

Write your food storage items in the "Item" column. Write the amount you need for a three-month supply in the "Need to Purchase" column. Each time you purchase the item, enter the quantity and the date purchased.

NEED TO PURCHASE	ITEM	QTY	DATE	QTY	DATE
10 pkgs	Oatmeal				
20 lbs	Pancake mix				
2 boxes	Cream of Wheat				

NEED TO PURCHASE	ITEM	QTY	DATE	QTY	DATE
2 gal	Syrup				
12	Gravy packets				
12 lbs	Lean sausage				
1 #10 can	Biscuit mix				

Now it's your turn to start filling out your breakfast shopping list. Use the breakfast shopping list worksheet you downloaded from our Web site at http://www.notyourmothersfoodstorage.com. As you shop and put your groceries away, remember to fill in the quantity and date you purchased each item.

Enlist the help of family members in shopping and filling out your shopping list. It's a wonderful feeling to see how quickly your food storage will grow with consistent and wise purchases.

PLANNING YOUR LUNCH MEALS

What does your family like to eat for lunch? Canned soup, or maybe just sandwiches? Lunch can be as simple as opening a can of soup or making a peanut butter and jelly sandwich. In our family, lunch is often leftovers from dinner the night before. Whatever it is, use our formula to figure out how much of each item you will need to store. Below you will see a sample lunch plan that we have included to give you some ideas. Look it over and then, on your own lunch meal planning worksheet, make a list of the lunch meals you want to have in your three-month food storage plan. For most families, lunch is just the bridge between a hearty breakfast and dinner. Keep your lunch meals simple.

Lunches	# times per month	× 3 months	× number in family	= total number of servings
Chicken noodle soup (canned)	4	12	4	48 servings
Tuna on crackers or bread	4	12	4	48 servings
Canned pasta (e.g., SpaghettiOs)	4	12	4	48 servings

Lunches	# times per month	× 3 months	× number in family	= total number of servings
Tomato soup (canned)	4	12	4	48 servings
Chicken salad on crackers or bread	4	12	4	48 servings
Bean with bacon soup (canned)	4	12	4	48 servings
Split pea soup (canned)	3	9	4	36 servings
Grilled cheese sandwiches	4	12	4	48 servings
Total meals	31			

Can you see how easy it is to plan your food storage by the meal? Why waste money buying more of an item than you'll need when that money could go toward another food item? In the sample lunch plan above, we kept it simple by rotating eight lunches and serving each of them four times, except for the split pea soup. We could have planned to rotate seven meals, or even four meals. It's easy to know just how much of any one item you need to store when you plan your food storage by the meal.

WRITE DOWN THE PLAN

Write down as many simple lunch meals as desired on your worksheet. Planning meals can be a fun family activity. Children are much more willing to go along with a plan in which they are a part. If you can't think of a different lunch for each day of the week, repeat a favorite lunch as needed so that you planned a lunch for each day of the week, with a total of thirty or thirty-one lunches for one month.

CREATE A LUNCH SHOPPING LIST

Creating your lunch shopping list should be easy if you've kept the meals simple. We'll show you how we created our shopping list for the lunch meals shown in our sample plan.

Canned soup: For canned soup, we figure that one can of soup, condensed type or not, will feed one teenager or adult, or two children. So, for our chicken noodle soup lunches, for example, we would need to purchase enough cans of soup for forty-eight servings. There are twenty-four cans in a case of soup, so we would write forty-eight cans or two cases on our shopping list, because every member of our family would eat a full can of soup. Then we would plan to buy one or more cases at a case-lot sale, or just pick up some extra cans when it goes on sale. At the last case-lot sale we attended we were able to get Campbell's chicken noodle soup for fifty cents a can, or twelve dollars per case.

Don't forget saltine crackers for soup, or the little oyster crackers, if you prefer. You'll have to decide how many crackers you'll allow each family member for each soup meal, or you can just store as many boxes as you can and allow them all they want. You'll find that saltine crackers are used in some of the recipes in our recipe chapter, so having extra on hand would probably be a good idea.

On the saltine box, it states that a serving is five crackers. If you want to use more crackers per serving, such as ten, just cut the total number of servings stated on the package in half and then decide how many boxes you need to store.

Tuna sandwiches: We figure that one can of tuna makes two sandwiches for each member in our family. So, for our tuna sandwich lunches we would need to purchase twenty-four cans of tuna. There are forty-eight cans of tuna in a case, so on our shopping list, we'll write twenty-four cans, or one case, which would give us extra tuna to use for one of our dinners. Don't forget to allow for mayonnaise or other condiments you like to mix in with your tuna. Put these on your staples/condiments shopping list.

Chicken salad sandwiches: A 13-ounce can of solid-pack chicken will probably make four sandwiches once mixed with mayonnaise or salad dressing. Use that estimate to determine how many cans you'd need to make forty-eight sandwiches. As with the tuna sandwiches, don't forget to consider condiments.

Grilled cheese sandwiches: For grilled cheese sandwiches, consider Velveeta or a store brand of processed cheese, which has a long shelf life. A 2-pound loaf of processed cheese may have a best-by date eight months out and requires no refrigeration until after it is opened. If you are using and replacing your three-month supply of food storage, you will be able to use both processed cheese and a number of other cheeses. For example, I recently purchased some shredded Mozzarella cheese with a best-by date five months out. I could easily store this cheese unopened in my fridge for three months.

When we make grilled cheese sandwiches with Velveeta, we use two slices per sandwich, which means we can make about sixteen sandwiches per 2-pound loaf. To make the forty-eight sandwiches called for in our sample meal plan, we'd need three loaves of Velveeta.

Canned pasta: One can of pasta will feed one adult or teenager, or two children. Because we have only teenagers left at home, we would need to purchase forty-eight cans of pasta for our three-month supply of these meals. These items come twenty-four to a case, so on our shopping list we would write two cases of canned pasta, probably SpaghettiOs and Chef Boyardee Beef Ravioli, or something similar. In our area, SpaghettiOs go on sale for about fifty cents a can, or twelve dollars per case. The ravioli are more expensive because they have meat in them; but we will buy them when we can get the best price because our teenagers love them. You don't have to buy a full case of one kind of canned pasta. You can mix the brands and just buy the number of cans needed when they go on sale so that you have more of a variety.

Bread, crackers, or tortillas for your sandwiches: If you can't get to the store to buy bread, or if the store shelves are bare, you've got a choice: you can either use crackers or tortillas in place of bread, or you can make your own bread for sandwiches. Trust me, knowing how to make bread will come in handy. It's easy to make, and it's a skill we would recommend to everyone, even dads and teens. Of course you can keep store-bought bread in the freezer, but it takes a lot of space. We've included recipes for you to use to make your own bread and your own tortillas. Another alternative to bread is pancakes. One of our neighbors often used leftover pancake batter to make extra pancakes and then stored them in the freezer. Whenever she ran out of bread, she used pancakes for their sandwiches. Their children loved them.

To make your own bread, you can store white flour right in the package it comes in for a three-month supply. For longer-term storage, check out chapter 9, How to Store It Now That You Have It, or store whole-grain wheat. For more information on storing wheat and how to use it, see chapter 6, Storing Staples and Condiments.

The shopping list for our sample lunch meals would look something like this:

LUNCH MASTER SHOPPING LIST

Write your food storage items in the "Item" column. Write the amount you need for a three-month supply in the "Need to Purchase" column. Each time you purchase the item, enter the quantity and the date purchased.

NEED TO PURCHASE	ITEM	QTY	DATE	QTY	DATE
2 cases	Chicken noodle soup				
2 cases	Bean with bacon soup				
2 cases	Tomato soup				
36 cans	Split pea soup				
3 boxes	Processed cheese loaf				
12 cans	Canned chicken				
2 cases	Canned pasta				

NEED TO PURCHASE	ITEM	QTY	DATE	QTY	DATE
15 boxes	Saltine crackers				
1 case	Canned tuna				

Now it's your turn to start filling out your lunch shopping list. Use the lunch shopping list worksheet you downloaded from our Web site at http://www.notyourmothersfoodstorage.com. As you shop and put your groceries away, remember to fill in the quantity and purchase date for each item.

Of course, you will be continually adding items to the staples and condiments shopping list that go along with your meals. For example, you could put peanut butter for PB&J sandwiches on your lunch shopping list or on your staples and condiments shopping list, depending on how much peanut butter you use. It may be considered a staple in some families.

If you've already completed your breakfast shopping list, filling out your shopping list for lunches should be even easier. Whenever you're scanning the grocery ads for specials, or going online looking for coupons, keep your shopping lists handy so that you never have to pay full price.

Chapter 4

PLANNING YOUR DINNER MEALS

When planning your dinner meals, it helps to brainstorm a bit before you actually make a menu. Here are some brainstorming tips:

- Make a list of ideas for the main entrée. Think about whether each entrée can stand alone—like a hearty homemade soup or a casserole that includes meats and veggies—or if it needs accompaniments, such as rice, potatoes, pasta, and/or a vegetable.

- Strive for variety. One way to ensure that your menus include a variety of foods is by planning to serve at least one meal with soup as the main course, one using pasta, one using potatoes, one using rice, and one using beans during any given week.

- Plan for miscellaneous meals that may or may not contain any of the above. These might be dishes like tamale pie or chicken enchilada casserole. Think of a variety of meats, poultry, and fish or seafood your family likes. Listing meal ideas in terms of ingredients will help you pick a menu that is varied and more nutritious.

Once you've listed a number of dinner ideas, look at each and evaluate how much your family likes that dish and how easily the recipe could be adapted to use storable items. After you've examined the list, make your meal plan by choosing fifteen to twenty dishes that you'll serve once or twice a month. The main dishes in the sample dinner plan included here can all be made from storable foods. We've included the recipe for each dish in chapter 11. For the sake of simplicity, our sample plan includes only the main entrée; but you can plan for additional side dishes that your family likes. Your dinner meal plan will look different than your breakfast and lunch meal plans because one main dish will serve the entire family. So, instead of thinking in terms of servings, most of the time, for dinner, you will be thinking in terms of how many times you will serve each meal in a month. Our sample dinner meal plan looks like this:

Dinners	# times per month	× 3 months = number of times meal is served
Spaghetti w/meat sauce	2	6
Macaroni and cheese	2	6
Fettuccine Alfredo w/chicken	2	6
Beef-barley soup	2	6
Salmon loaf	1	3
Chicken corn chowder	2	6
Clam chowder	2	6
Chicken enchilada casserole	2	6
Tacos and beans (canned)	2	6
Meat and gravy over mashed potatoes or rice	2	6
Spanish rice w/chicken	2	6
Chili and crackers	2	6
Hamburger Helper (your choice)	2	6
Creamed tuna on toast or biscuits	2	6
Cheesy potatoes with ham	2	6

Dinners	# times per month	× 3 months = number of times meal is served
Tamale pie	2	6
Total Meals	31	

BEGIN YOUR DINNER PLAN NOW

To create your own plan for dinner meals, use the dinner meal plan worksheet to list the meals you have chosen. You can list side dishes, too, if you want. It's up to you. If you plan on having canned or freeze-dried fruits or vegetables on hand to go with your main dishes, use the worksheet to figure out how much to store, just as you did for breakfast and lunch meals. Put each of your side dishes—or fruits or vegetables—on a separate line so that you can more easily figure out just how much of each item you'll need to store.

Get the family involved and come up with as many meals as you want in your food storage. It's helpful if you can plan to use some food items for more than one meal so that you don't have to shop for so many foods. For example, if you plan on chicken for two or three meals, add the quantities together when you create your shopping list. Whatever you decide to store, be sure it's something your family will actually eat, because you will be using it to prepare meals and then replacing the items when they go on sale.

YOUR SHOPPING LIST FOR DINNERS

You can get a little more creative with dinners if you want. Unless you've planned for Hamburger Helper-type meals, you'll be looking at your recipes to figure out how much to buy of each ingredient. So, on your shopping list for dinners you may be listing ingredients when planning a meal made from a recipe, or you may list a product, such as Hamburger Helper, or canned vegetables or fruits for side dishes. For the sake of simplicity, we'll confine our sample shopping list to just the main dish for each meal so that you can see how it's done. From there you can expand and add whatever side dishes you want. Work through one meal at a time to determine exactly what you'll need to buy to prepare it.

Spaghetti: The first meal on our sample worksheet is spaghetti with meat sauce. Storable pasta dinners usually consist of canned sauce, pasta, and meat. You can choose a sauce that has meat in it, or you can always choose spaghetti sauce without meat. We decided that we would have spaghetti sauce twice each month, which means we would need to store enough sauce, spaghetti noodles, and meat for six meals.

You can buy pasta in bulk at ten pounds per package, or you can buy it during case-lot sales, and/or use coupons to keep from paying full price. On average, one pound of pasta will yield eight servings. Using that measurement and knowing that we need to prepare six meals, our shopping list would show three pounds of spaghetti noodles. If your family likes larger portions, consider adding another pound to your shopping list.

We would need one can or jar of spaghetti sauce for each meal, which means we would put six jars of spaghetti sauce on our shopping list. You can buy spaghetti sauce with meat included in the sauce, or you can add ground beef or textured vegetable protein (TVP) to a meatless sauce. If you want to add meat to your sauce, plan to store one pound per meal, which for the sample plan would equal six pounds of ground beef or the amount of TVP that will equal six pounds when reconstituted. Of course, you can always opt to make your spaghetti sauce from scratch, so we've included a simple recipe in chapter 11.

Macaroni and cheese: For macaroni and cheese dinners, you can use a package mix, which is easy to store and economical. Check the number of servings on the package and make sure that the serving size is large enough for each member of the family. In our sample plan for dinner meals, we plan to have macaroni and cheese twice each month, and we usually make two boxes each time. So, for our family we would add twelve boxes of macaroni and cheese to our shopping list. Or, we could choose to make our own macaroni and cheese and store the needed ingredients: macaroni noodles and processed or—for shorter-term storage—shredded cheese. An easy recipe for macaroni and cheese is included in chapter 11.

Fettuccine Alfredo with chicken: The third meal in our sample plan is Fettuccine Alfredo. We also plan to serve this twice each month. To know how much fettuccine to store, check the package to see how many servings it makes and then use the formula in the worksheet to arrive at the number of servings. A 12-ounce package contains six 2-ounce servings. We don't think two ounces will do it for our family, so we would plan on 4 ounces instead. In that case, we know that we need to buy eight 12-ounce boxes of fettuccine (twenty-four servings divided by three servings per box).

We could use canned chicken for Fettuccine Alfredo and add frozen broccoli and/or cauliflower to turn it into a pasta-primavera type dish. It takes one can of chicken for each meal, so we would add six cans of chicken to our shopping list, plus any vegetables we plan to serve with it.

For the Alfredo sauce, you can buy it or make it from scratch. We plan to buy the sauce rather than try to store that much grated Parmesan cheese. We would need six jars or twelve packets of Alfredo sauce, because we plan to serve this meal six times during a three-month period. If we wanted to make the sauce, we would check our recipe and then figure how much we would need of each ingredient for six meals (we'll walk you through that process when discussing our Beef Barley Soup recipe). We would then put these items on our shopping list.

Beef-barley soup: The next meal on our sample dinner worksheet is beef barley soup, one of our favorites. For any soup, you will need to figure out how much of each ingredient to store. We plan to serve this soup twice each month. For this meal, we'll need approximately one pound of ground beef, canned beef, or enough TVP or bulgur to reconstitute to equal one pound of ground beef for each pot of soup. We can use fresh (if available), canned, freeze-dried, or dehydrated vegetables, quick or regular barley, and shell pasta. Refer to the recipe, which makes eight servings and then multiply the ingredients to make the soup six times according to our sample three-month plan. The nice thing about soup is that there is often leftover soup that can be served the next day for another meal. Let's walk through this process so that you can see just how easy it is.

Here is the list of ingredients needed for beef-barley soup, a hearty soup that our family loves:

Beef-Barley Soup

1 pound lean ground beef
1 onion, chopped
8 to 10 cups water
8 to 10 teaspoons beef bouillon
½ teaspoon onion powder
½ teaspoon garlic powder
Salt and pepper to taste
¾ cup barley
4 or 5 carrots, sliced (or 1 can sliced carrots, or 2 cups reconstituted
   freeze-dried carrots)
1 cup medium shell pasta, uncooked

To figure out what to put on your shopping list for soup, or any meal for which you use a recipe, first decide if you will need to substitute any ingredients for ones that wouldn't store well. We'll go through the ingredients one at a time.

Ground beef: One pound is needed for each pot of soup. Put six pounds on your shopping list.

Onion: Will you store fresh onions or dried onions? This item would go on your staples and condiments shopping list, as would all of the seasonings in this recipe, including the beef bouillon. If you want to store whole onions, check out the tip on storing onions in chapter 9, How to Store It Now That You Have It.

Seasonings: These items—like onion and garlic powder, and salt and pepper—will go on your staples/condiments shopping list because you'll use them in so many other meals.

For other soup recipes, follow the same general method: refer to the recipe for one meal and multiply the ingredients to equal the number of times you will serve the soup. You can also store commercial soup starters for your homemade soups. Make sure that you store both chicken and beef bouillon as well as dried minced or chopped onions and celery seed. With onions, celery, carrots, and bouillon, you can make any soup taste delicious, even if fresh ingredients are not available. I enjoy experimenting with tomato bouillon, too.

Barley: Will you buy boxes of quick barley or buy barley in bulk? We use the quick barley, which cooks up in ten minutes. Storing barley in bulk is much cheaper, but it takes longer to cook. If you decide on the boxes of quick barley, the 11-ounce box of Quaker Quick Barley will make about three pots of soup, so you would add two boxes of quick barley to your shopping list.

Carrots: Will you have fresh carrots from your garden (which will store in the ground for several months, depending on where you live), or will you use freeze-dried or canned carrots? It takes four or five carrots for the soup, which would slice up to about 2 cups of sliced carrots. For the sake of simplicity, we'll use canned carrots, so we'll add six cans of carrots to our shopping list because we plan to have this meal twice each month. (As you find other meals for which you'll need carrots, the number for carrots may grow.)

Medium shell pasta: A 1-pound box of medium shells will give you 6 cups of dry pasta. It takes 1 cup per pot of soup, and we're serving the soup six times, so one box is all we will need. However, if we use these pasta shells in other recipes, we may put more boxes on our shopping list.

Vegetables: Decide what kinds of vegetables you will store for your meals: canned, frozen, freeze-dried, or a combination of all three. Canned whole new potatoes can be used for soup whenever fresh potatoes are not available, and dehydrated hash browns can be reconstituted and cook up nicely for a potato-based soup.

Hamburger Helper: Jan's family loves several of the varieties of Hamburger Helper, and it's a quick meal when time is short for preparation. We're serving this six times during the three-month period, so on our shopping list we would write six boxes of Hamburger Helper (different varieties). Next we would add six more pounds of ground beef to our shopping list.

Are you getting the hang of this? It's really a simple concept, but having your meals planned will make all the difference when you are faced with having to live off of your food storage for a period of time— or even any time your month outlasts your money. Just have a delicious food storage meal that you know your family will love.

Tacos and beans: Our family loves tacos, especially when made with our own homemade tortillas. Making tortillas is surprisingly simple and quick. And fresh tortillas taste great. When serving tacos while living off your food storage, you may not have fresh lettuce or tomatoes on hand unless you have a garden or you are able to reserve a little cash for fresh produce. (For a tip on how to store tomatoes from your garden for several months, see chapter 9, How to Store It Now That You Have It.) Experiment with ingredients that you have on hand when fresh items aren't available.

Our shopping list for this meal would consist of *masa harina*, the corn flour used to make corn tortillas, refried beans or chili, canned diced tomatoes or salsa, olives, ground beef, and cheese. (We include recipes for corn and flour tortillas in chapter 11.) Since grated cheese stores well in your freezer, you should be able to have this as part of your food storage. We're including a recipe for a delicious recipe of Mexican meat that you can make and freeze. A quick and easy Spanish rice recipe is also included in our recipe section to round out a taco meal.

The 4.4 pound package of *masa harina* will make 128 six-inch tortillas, so one package will probably be plenty for a three-month period. You just add water to the mix and follow the directions on the package.

Whatever else you like to put on your tacos, just multiply it by the times you want to serve tacos and that's what you put on your shopping list.

For our shopping list, we'll add enough packaged taco seasoning, canned chili, and canned diced tomatoes and olives for six taco meals. Adding six pounds of ground beef to what we already have on the shopping list would bring the total up to eighteen pounds. To our staples/condiments shopping list, we'll add the olives and hot sauce. We'll be using olives for other recipes, as well as for snacking.

Chili: You can buy canned chili, which is what we will do for our shopping list, but making homemade chili is quick and easy. We've included a recipe in chapter 11. If we use canned chili, it takes three cans for a meal of chili, with corn bread or crackers, so for six meals of chili, we would add eighteen cans of chili to our shopping list.

Fruits and vegetables: Although we haven't shown them in the sample worksheet, be sure to store whatever fruits and vegetables your family likes to eat. You can store frozen, freeze-dried, or canned fruits and vegetables, depending on your budget and freezer space. Not only do they round out a meal but fruits and vegetables also add needed nutrients to your diet. With spaghetti, in place of a fresh salad, we like canned green beans. For our family, it takes two cans, so we would add twelve cans of green beans to our shopping list. As you fill out your own dinner meal plan, decide what fruit or vegetable might go well with each meal.

So far, our shopping list for just our main entrées would look like this:

DINNER MASTER SHOPPING LIST

Write your food storage items in the "Item" column. Write the amount you need for a three-month supply in the "Need to Purchase" column. Each time you purchase the item, enter the quantity and the date purchased.

NEED TO PURCHASE	ITEM	QTY	DATE	QTY	DATE
3 lbs	Spaghetti noodles				
6 jars	Spaghetti sauce				
18 lbs	Ground beef				
6 lbs	Fettuccine noodles				
6 jars	Alfredo sauce				
2 boxes	Quaker quick barley				
2 boxes	Medium shell pasta				
6 boxes	Hamburger Helper				
12 boxes	Macaroni and cheese				
6 cans	Carrots				
6 cans	Chicken				
1 pkg	Masa harina				
18 cans	Chili				

Use the same methods and formulas we've outlined in previous chapters to plan your shopping list for dinner meals. Simple meals work best. In times of plenty, supplement with other items to make your meals more gourmet. In hard times, use as is and enjoy a full tummy. If you plan dinners that your family loves to eat, you will be able to practice the "use and replace" method of rotating and maintaining your food storage. The big plus for this method of acquiring your food storage is that when times are tough your family won't have to deal with a sudden change in their diet. That means less stress on everyone.

Use the dinner shopping list worksheet you downloaded from our Web site at http://www.notyourmothersfoodstorage.com. As you shop and put your groceries away, remember to fill in the quantity and purchase date for each item.

Chapter 5

PLANNING FOR DESSERTS AND TREATS

Don't forget to store items for an occasional dessert or treat. When your family is trying to adjust to any kind of crisis, sweets can go a long way in lifting everyone's mood. A can of pumpkin, a spice cake mix, and a bag of chocolate chips make a batch of great-tasting cookies. If you have fresh apples, baked apples can be a wonderful treat. Jell-O and puddings are other treats that store well. Cake and brownie mixes store quite well also, especially if you are practicing "use and replace." Popcorn is a great treat and will store for a few months. One of my grandsons loves eating graham crackers that have been broken up and served with milk in a bowl. Other grandchildren love graham crackers with canned or homemade frosting spread on top.

Think about your favorite desserts or snacks and which ones can be made with ingredients that are storable. Then just use our formula to create your shopping list for three months. Here are some desserts and treats to consider. Every treat can be made from items that store well. If you want ice cream, you'd have to have enough room in your freezer, or plan to make your own, which would probably be a lot more fun.

Desserts and Treats	# times per month	× 3 months = number of times dessert is served
S'mores	2	6
Pumpkin cookies (use recipe)	2	6
Jell-O	2	6
Pudding	2	6
Popcorn	3	9
Ice cream	3	9

Now that you've planned your desserts and treats, make a shopping list just like you did for your dinner meals. You won't be planning for servings; instead you'll be planning for how many times each month you want to serve a particular treat. You don't have to worry about having a total of thirty or thirty-one desserts unless you want to. Jan and I have planned for occasional treats, preferring to spend our money on foods that we can use to create a meal. The choice is yours.

When making your shopping list for treats, you can use a separate shopping list or just add the items to your dinners shopping list. Both can be downloaded from our Web site at http://www.notyourmothers foodstorage.com.

STORING STAPLES AND CONDIMENTS

Staples are the items you never want to be without. This list will vary depending on what foods you like to prepare. For most people, staples include items like flour, salt, sugar, and oil. The best way to figure out what staples you need to store is to look in your cupboards and write down a list of everything you use regularly. Include spices and condiments on the list. You could also make a list of things you use occasionally or only on special occasions and pick up some of those items for your storage as well. I've found that the spices and seasonings I use most are seasoning salt, onion powder, garlic powder, cinnamon, vanilla, Italian seasoning, and bouillons, such as chicken, beef, and tomato. I also use dried onions, either minced or chopped, in some recipes. You can buy large sizes of most spices at wholesale shopping clubs or you can just watch for sales. Once a month or so, pick up a few extra of the items on your staples list.

Many staples and condiments store quite well. Salad dressing, mayonnaise, ketchup, and mustard can often be stored for a year in a cool place. These are among the items you'll want to be sure that you use and replace so that they don't sit in your storage area for several years. As always, check for a best-by date and then adjust the amount you buy accordingly.

What follows is a discussion of common staples and things you should keep in mind when storing them.

Oil: People often don't store enough oil. During World War II, oil was one of the rationed items most in demand. Olive oil stores better than all other oils, lasting up to two years if kept in a cool, dark place. Canola oil is an excellent oil to store because it is very low in saturated fat and high in monounsaturated fat. Consider storing olive oil for longer-term use and canola and vegetable oil for daily use. Canola and vegetable oil will store well for up to a year if kept at a cool temperature.

Wheat: If you decide to store wheat, we recommend hard white wheat. Because of its milder flavor and lighter color, it's much easier to get used to eating. In addition to hard white wheat for bread, there is soft white wheat for other kinds of baking, such as sweet breads, cookies, and pancakes. This grain will store for about eight years (while hard wheat will store indefinitely in the right conditions) and is delicious and wholesome when used in recipes.

If you choose to store wheat, you'll need to invest in a grain mill, unless you have a friend or family member who has one and will allow you to use it. Electric models are priced around $200, and hand models sell for just over $100 for the good ones. However, you can get an electric model for about $150 when you buy several at a time from BlendTec, which our ward has done several times. Families can also split the cost of a grain mill and share the use of it.

BlendTec is a good resource for both types of mills: http://www.blendtec.com/Mills.aspx. They give quantity discounts for families who want to order at the same time. These are excellent mills and will grind wheat, corn, beans, and more. I have used my electric mill from this company since 1992 and it's great. Another excellent brand is the WonderMill, which is slightly more expensive but uses stones instead of blades and is quieter to run. Their Web site is http://www.thewondermill.com.

Other items you'll want to store to make your own bread are: SAF instant yeast, oil, honey, and salt. I also add gluten and dough enhancer for my bread to make it lighter and to increase the protein content and shelf life. A fool-proof recipe for delicious homemade whole wheat bread can be found in chapter 11. All of these items are available at grocery stores that sell food storage items, or places like Bosch kitchen stores or health food stores. In our area, Walmart now carries the yeast, gluten, and dough enhancer. You can also order these items online. Two good sources for gluten, SAF yeast, and dough enhancer, as well as other baking products, are http://www.pleasanthillgrain.com/grains_baking_supplies.aspx or http://www.augasonfarms.com/.

Milk: Milk is another essential item, for drinking and for recipes. Your family may be like ours and turn up their noses at powdered milk. However, there were times when that was all I had, so I had to get sneaky. I rinsed out the milk carton, mixed up powdered milk, and put it back in the carton the night before so that it could chill all night. The next morning when the children poured milk on their cereal, I didn't hear a peep (most of the time). It worked even better if there was a little milk left in the carton. I would then just add the reconstituted powdered milk to the milk left in the carton. Jan often does the same thing when she runs out of milk. For drinking, you can add a little chocolate milk powder if that's the only way you can get them to drink milk.

Whey-based milk substitute tastes better than straight powdered milk but is not as versatile. This type of milk product is sold under the brand name Morning Moo's, among others. With regular powdered milk you can make cheese, yogurt, and more, should you decide to do so. Another alternative to milk is canned evaporated milk, which can be used in most recipes if diluted by half with water. It is inexpensive and easy to store.

Shelf-stable milk is another alternative used widely in Europe. I recently purchased three cases of shelf-stable milk: two cases of 2 percent and one case of half and half. You can also get whole milk, skim milk, whipping cream, and chocolate milk in the shelf-stable cartons from several sources. Shelf-stable milk is ultra-pasteurized and requires no refrigeration until it is opened. It tastes just like the milk you buy at the store. It has no preservatives and all of the same nutrients and will store for several months at room temperature. I bought mine from a local source (http://www.alpinefoodstorage.com), but it came from Gossner Foods. You can read more about it on their Web site (http://www.gossner.com) or by doing an Internet search for shelf-stable milk.

Although milk can be frozen, it takes up a lot of freezer space. How much milk does your family go through in a week? Multiply that number by twelve weeks and store enough powdered milk to match your family's needs for milk. If you mix it ahead and then chill it before serving, it tastes pretty good and your family will get used to it when they know there is nothing else. Or store the shelf-stable milk, but if you do, you'll want to use it along with your regular milk and then periodically replace it.

Eggs: You'll want to keep enough eggs on hand to last you five or six weeks. They'll keep that long in the fridge. Powdered eggs can be purchased in several varieties, such as whole eggs, egg whites, egg yolks, and scrambled egg mix. If you choose not to use powdered eggs, consider doing what we have found to work well: beat raw eggs with a fork in a bowl and then store them in freezer bags in the freezer. We put two eggs in each quart-sized freezer bag and store them lying flat in the freezer. To use, just thaw under cool water and use for scrambled eggs, German pancakes, or in any baking you plan to do. They take up hardly any space and will keep well for three months. Be sure to write the date that you freeze them on the package.

Another item to consider in place of eggs is unflavored gelatin. In chapter 11 you'll find directions for making an egg substitute that uses unflavored gelatin and can be used in baked goods like brownies or cookies.

Other staples: Store white flour, sugar, honey, and salt in whatever quantities you can afford. Try budgeting a small amount each month for these staple items. Do the same with cooking oils. When you see baking items like chocolate chips on sale, usually just before

Thanksgiving and Christmas, pick up some extra bags. We've listed below some of the staples and condiments that we store and make sure we are never without.

- Baking powder
- Baking soda
- BBQ sauce
- Beans
- Beef bouillon
- Bulk onions
- Bulk potatoes
- Canned evaporated milk
- Chicken bouillon
- Chocolate chips
- Cinnamon
- Cooking oils (olive, canola, vegetable)
- Dough enhancer
- Dried onions (minced or chopped)
- Garlic powder
- Gluten
- Graham crackers
- Honey
- Italian seasoning
- Ketchup
- Maple flavoring
- Mayonnaise or salad dressing
- Mustard
- Onion powder
- Pepper
- Powdered milk
- Raisins
- Rice
- SAF yeast
- Salt
- Saltine crackers
- Seasoning salt
- Shortening
- Sugar
- Tomato bouillon
- Unflavored gelatin
- Vanilla
- Wheat
- White flour
- Worcestershire sauce

In addition to the list above, we store dehydrated hash browns and potato pearls (also sold by Harvest of the West as potato gems). Jan is able to store potatoes in her root cellar for short-term storage. I'm not so lucky, but I live close and she shares.

Also consider canned meats if you don't have much freezer space. Most stores carry canned ham, Spam in several varieties, canned roast beef and gravy, roast beef hash, and corned beef hash. And canned tuna, salmon, and chicken are available in all supermarkets.

BEGIN PLANNING STAPLES AND CONDIMENTS

Start now listing the items that you feel you can't be without. Look in your pantry and take stock of what you have on hand. Start keeping track of how much you use in a week and then multiply so that you know how much you need to store for a three-month supply. A worksheet like the one below is available at our Web site: http://www.notyourmothersfoodstorage.com.

Staples and Condiments	How much we use in 1 week	× 4 = 1-month supply	× 3 = 3-month supply	× 4 = 1-year supply

Some staples, like pepper, baking powder, baking soda, and seasonings seldom go on sale but can be purchased in large sizes at wholesale shopping clubs or online. I recently found a fabulous seasoned rub for beef and pork when I was visiting my family in California. When I returned home, I checked every store in our area but couldn't find the same rub. I went online, searched for the name of the product, and found a distributor and a great price. I am now well stocked on this product.

YOUR SHOPPING LIST FOR STAPLES AND CONDIMENTS

The items on your staples and condiments shopping list will be the ones that you always keep on hand. Once you have arrived at the amount your family uses in a month and have multiplied that by three to know what you need for a three-month supply, start filling out the staples/condiments shopping list. Transfer each item and amounts you have listed in the "×3 = 3-month supply" column to the "Need to Purchase" column on your shopping list.

Now you are ready to start using the methods we've mentioned earlier to save money on these important items. As you shop and put items away, make sure you fill in the quantity and date they were purchased. These are the items you never want to be without, so as you use them, be sure that you don't let your pantry get too low. You never know what tomorrow may bring. Be prepared.

Chapter 7

STORING NONFOOD ITEMS

Nonfood items are quite often overlooked in food storage plans. Imagine what you'd do, however, if you ran out of toothpaste, toilet paper, shampoo, deodorant, feminine products, paper towels, dish soap, dishwashing detergent, laundry detergent, bleach, fabric softener, cleaning solutions, and so on. Obviously, many of these items are nearly as vital as food. As you put together your list of nonfood items to store, stop and consider what you really cannot do without in an emergency. It seems like the things you can't eat always cost more, item by item, than the food you buy. Keep this list as small as possible, because if it's not food, and you can't eat it, do you really need it in an emergency?

It's simple to figure out how much to store of these items. Just do what Jan did: figure how much you use in a week or month and multiply to get the quantity you need for three months. If you're not sure how much you use of an item, buy a specific single item, such as dish soap, and write the date you purchased it on the label. When it's gone, check the date and you'll know how long one bottle of dish soap will last you. You could do the same with toilet paper or toothpaste, for example. From there it's simple math to find out how much you need to have on hand for a three-month supply.

If you find a really good deal on a product you use all the time, buy enough for three months or more. Or, pick up an extra item whenever possible. If you use the Savvy Shopper coupon system (or another coupon system), most nonfood items will cost very little and may even be free. For example, our friends who use this system religiously told us that they rarely pay anything for shampoo because they only buy it when it's on sale and when they have a coupon for that brand. They aren't picky about the brand and often will use shampoo as a body wash in the shower. Toothpaste is another item for which they say they never pay more than twenty-five cents.

Another way to save money on nonfood items is to consider using common household items for cleaning, such as hydrogen peroxide, white vinegar, baking soda, and lemon juice, which work incredibly well to remove stains, mildew, soap scum, and all kinds of household grime. For specific recipes go to http://www.EarthEasy.com or http://www .vinegartips.com/cleaning or do an Internet search for "homemade cleaning supplies."

Here is a list of some of the nonfood items that are typically used:

- All-purpose cleaners
- Baby supplies, if needed
- Cosmetics
- Dental floss
- Deodorant
- Dish soap
- Dryer sheets/fabric softener
- Feminine supplies
- First-aid items
- Laundry detergent
- Medicines
- Paper towels
- Shampoo/conditioner
- Toilet paper

- Toothpaste
- Vitamins
- Zip-top bags

When it comes to first-aid items, you can buy a kit or make your own. Jan made a fabulous first-aid kit for each of her married children from soft-side fishing tackle boxes that she purchased from Walmart for about $14.95. She bought most of the supplies to stock the kit at wholesale prices from http://www.statmedical.com. Several families went together to place a bulk order, dividing the cost and the products. To see a picture of the kit and get the free step-by-step instructions, including the labels for each partition, go to http://www.kathybraybargains.com. You can find many sources for first-aid items online by doing a simple Internet search. A first-aid kit isn't used only in times of emergency. It's there to be used anytime you need it, and when all of your most commonly used items are in one place, it saves time hunting for a bandage or bee-sting medication. Don't forget to store over-the-counter pain and cold relievers in your kit.

Begin your list of nonfood essentials on the worksheet you downloaded from http://www.notyourmothersfoodstorage.com. Make the list first, then start paying attention to how long essential items last in your family. That will give you the basis for figuring out how much it will take for a three-month supply of an item.

The worksheet below is just a sample of the full-sized worksheet that can be found online at our Web site.

Nonfood Essentials	How much we use in 1 week	× 4 = 1-month supply	× 3 = 3-month supply	Quantity	Date

Once you arrive at how much you'll need of a specific item, write it on the shopping list worksheet. Keep all of your shopping lists handy whenever you are scanning the newspaper for sales. The Sunday newspaper usually has one or two coupon inserts. We've found these are often filled with coupons for nonfood items. If you use the coupons when the items are on sale, you'll save a lot of money.

ADAPTING YOUR FAVORITE RECIPES

Your favorite recipes will probably need some adapting so you can make them from storable foods. Some recipes, such as a green salad or some vegetable stir-fry dishes, really won't adapt well if the main ingredient must be fresh. Cream cheese is also an ingredient that is not easy to substitute; but it has a long shelf life, so if you're careful to check the best-by date on the package, it should not be a problem to always have some on hand. If you use and replace sour cream, you should be able to have it on hand most of the time, as well. If you shop at large markets where they turn over a large quantity of products daily, you'll find the best-by date on sour cream is often six weeks or more from the purchase date. You can buy powdered buttermilk to use in recipes.

Most recipes are easily adapted. We'll show you how we do it and then you can apply the same steps to your own favorite recipes. Here is a sample:

Chicken Pot Pie

    2 pounds chicken or turkey, cooked and chopped
    1 (12-ounce package) frozen mixed vegetables, or vegetables of
        your choice

½ cup chicken broth (or broth from drained cans of chicken or turkey)

2 (10.75-ounce) cans cream of celery soup

½ cup butter, melted

2 cups biscuit mix

2 cups milk

Preheat oven to 350° F.

Layer cooked chicken or turkey in the bottom of a 9x13-inch casserole dish. Sprinkle frozen veggies on top. Combine broth and canned soup in a small bowl, then pour mixture over vegetables. Drizzle butter over the top. In a medium bowl, combine biscuit mix and milk; pour over top of casserole. Bake, uncovered, 1 hour.

To adapt this recipe, you would only need to substitute the fresh chicken for two cans of chicken. If you have frozen vegetables, use them; otherwise, substitute canned vegetables.

Let's try another recipe.

Spanish Rice and Chicken

3 cups water or chicken broth (if using broth, omit salt)

2 cups rice

2 cloves garlic, finely minced

4 green onions with stems, chopped

2 teaspoons salt

1 (8-ounce) can tomato sauce

2 cups cooked chicken or turkey pieces (leftovers work well)

Bring water to a boil over high heat. Add rice and other ingredients. Reduce heat, cover, and simmer for 20 minutes or until rice is done.

To adapt this recipe all you would need to do is substitute dried chopped chives or onions for the fresh onions, and canned chicken or turkey for the fresh chicken. Simple, but delicious. Another variation would be to use Knorr Tomato Bouillon and omit the chicken broth, tomato sauce, and salt. Just increase the water to 4 cups and add 4 tablespoons tomato bouillon.

Here's another favorite recipe that has a few more ingredients:

Beef Stew and Dumplings

1½ pounds lean stew meat

Flour, for dredging meat

2 tablespoons vegetable oil

6 cups hot water

1 onion, diced

3 stalks celery, chopped

2 bay leaves

3 teaspoons beef bouillon

Salt and pepper to taste

8 medium potatoes, chopped

8 carrots, chopped

2 cups biscuit mix

⅔ cup milk

Dredge stew meat in flour to coat. Heat oil in a 6-quart pot on medium-high heat. Brown stew meat. Add extra flour, if needed, to absorb any extra oil. Pour in hot water and stir, making a thin, brown gravy. Add onion and celery to meat mixture. Break bay leaves into 2 or 3 pieces and add. Add the beef bouillon and salt and pepper to taste. (I use a seasoning salt.)

Reduce heat and simmer over low heat for 2 hours, stirring occasionally to keep meat from sticking to the bottom of the pot. Add chopped potatoes and carrots. Cook for 10 minutes.

Prepare dumplings by combining biscuit mix and milk with a fork. Drop by spoonfuls on top of simmering stew. Cook 10 minutes covered and another 10 minutes uncovered. Spoon dumplings onto a platter. If the gravy needs to be thickened, stir in a mixture of cold water and flour. Mix well and pour into gravy as you stir. Serves 6 or more.

To adapt this recipe, use whatever frozen beef you have on hand, even hamburger will work. If you don't have any frozen beef, use canned meat, such as canned roast beef and gravy. Substitute canned whole new potatoes for fresh potatoes, substitute canned carrots for fresh carrots, and use dried chopped onions. For the dumplings, if you don't have a biscuit mix in your food storage you can make them from scratch, which would take flour, baking powder, salt, and shortening, all of which you should have on hand.

The main thing to remember when adapting recipes is to look at the recipe ingredients and ask yourself, "What ingredient(s) do I need a substitute for, and what do I have on hand that I can use?" Although the finished meal may not be exactly the same as when made with the familiar fresh ingredients, it will still be delicious and nourishing.

Here is a handy chart for common ingredient substitutions:

Ingredient	Amount	Substitution
Allspice	1 teaspoon	½ teaspoon cinnamon plus ¼ teaspoon ginger and ¼ teaspoon cloves
Buttermilk	1 cup	1 cup plain yogurt OR 1 tablespoon lemon juice or vinegar plus enough milk to make 1 cup
Cornstarch, for thickening	1 tablespoon	2 tablespoons flour (approximately)
Egg	1 egg	1 teaspoon unflavored gelatin in 2 tablespoons boiling water. Whisk to dissolve. Mix in 2 tablespoons water.
Egg	1 egg	2½ tablespoons powdered egg substitute plus 2½ tablespoons water
Egg	1 egg	3 tablespoons mayonnaise (not light)
Fats for baking	1 cup	1 cup applesauce
Garlic	1 clove	⅛ teaspoon garlic powder
Ketchup	1 cup	1 cup tomato sauce plus 1 teaspoon vinegar and 1 tablespoon sugar
Sour cream	1 cup	1 cup plain yogurt OR 1 tablespoon lemon juice or vinegar plus enough cream to make 1 cup OR ¾ cup buttermilk mixed with ⅓ cup butter
Unsweetened chocolate	1 square (1 ounce)	¾ tablespoon cocoa plus 1 tablespoon shortening

For a list of additional substitutions, go to: http://allrecipes.com/HowTo/Common-Substitutions/Detail.aspx.

HOW TO STORE IT NOW THAT YOU HAVE IT

There are some fresh foods that you'll just want to have on hand whenever possible. Jan and I have learned some neat methods for storing some of these foods.

STORING FROZEN MEATS

Don't store hamburger or other meats in the package they came in from the market. Instead, put meat in a quart- or gallon-sized zip-top bag, making sure to get out as much of the air as possible. Press-and-seal freezer plastic wrap works well as long as you get out all of the air and carefully seal around the edges of the meat. Even better than that would be to purchase a vacuum-sealing unit. Meat that is vacuum sealed will maintain both color and taste. I have eaten year-old meat that was vacuum packed, and it was delicious. I have also sampled "butcher-wrapped" meat after it had been in the freezer for more than six months. It was freezer burned and tasteless. We couldn't eat it. Plan carefully what you freeze and how to wrap it so that nothing goes to waste.

STORING/PRESERVING FRESH PRODUCE

If you have the space and time, gardening and preserving your harvest is a wonderfully rewarding family activity. Most fruits and vegetables can be preserved by canning, but be sure to carefully follow prescribed methods for each kind of food. Those with high acidity, such as tomatoes and fruits, can be preserved using a hot water bath. For food safety, low-acid foods, such as green beans, must be preserved using a pressure cooker. Consider sharing the cost of a pressure cooker with friends and then sharing the use.

Tomatoes: Tomatoes are easy to can. You can use tomatoes to make stewed or diced tomatoes, salsa, and spaghetti sauce. There is some cost initially when you buy your bottles and lids, but you can reuse the jars and will only need to buy lids.

The best book we've found is the *Ball Blue Book Guide to Preserving*, which gives complete instructions for preserving food. It covers home canning, freezing, and dehydration from the basics to advanced techniques. It also includes recipes. I found it online for only $6.99 at http://www.allamericancooker.com/bluebook.htm.

Fresh onions: If you prefer using fresh onions, like we do, consider storing whole onions by using a new, or freshly laundered, pair of panty hose. Cut the hose apart so the legs are separate. Drop an onion into the toe of one leg of the panty hose. Tie a knot just above the onion and then drop another onion into the panty hose leg. Repeat this process until the leg is full and then do the same with the other leg of the panty hose. Hang in a dark, very cool place (the ideal temperature is around 50 degrees F.) such as a pantry, closet, or cellar. When you want a fresh onion, simply cut just below the knot to release the onion in the toe of the panty hose. Continue in this manner whenever you want a fresh onion for your recipes. If you grow your own onions, you can just braid the tops and hang to dry.

Potatoes: Potatoes will keep in a cool, dry place, but if you want to store them over the winter, consider using a garbage-type can and layering the potatoes with sand. Store them where they will not freeze.

Green tomatoes: Where we live, the frost comes when there are still many luscious, green tomatoes on the vine. I've used this method to store tomatoes for over two months. Choose only green tomatoes from healthy, vigorous vines without cracks or holes of any kind in them. Remove the stems, wash in cool water, and dry thoroughly with paper towels. Wrap each tomato in newspaper and store in layers no more than two or three deep in a wooden crate or open cardboard box. Store in a cool, dry place, with the tomatoes that are ready to turn first on the top layer. Check tomatoes every few days. As they become pink, put them on the counter to continue ripening. When ripe, I try to use them as soon as possible. If I have to keep them several days before using, I store them in the refrigerator, although they don't taste as good as those kept at room temperature. You can find other ways to store tomatoes by doing an Internet search on storing green tomatoes.

Peppers: Whether I grow them or just buy a bunch when they are inexpensive and in season, I like to dice and freeze them in small freezer bags. Then, whenever I have a recipe that calls for green pepper, I've got it on hand. I like to store diced peppers by the cup in freezer bags so that I don't have to thaw out a large package for just a small amount. Frozen, diced green peppers are also available in the freezer section of most grocery stores.

Zucchini squash: I like to grate zucchini squash when it's in season and store it in the freezer in 2-cup amounts. This is the amount needed for zucchini bread or zucchini chocolate cake. Grated zucchini is also delicious in meatloaf and soups.

Root crops from your garden: Root crops, such as carrots and beets, can be left in the ground. A heavy mulch of straw will help prevent the ground from freezing so the vegetables can be dug up when needed. The mulch reduces the freeze-and-thaw cycle, which can reduce the quality of the vegetables left in the ground. Root crops stored in this manner become sweeter and milder. If temperatures drop so low that the ground freezes below the mulch for several inches, it's time to harvest. Cut off all but about a half inch of the leafy top before storing. Store in cool temperatures (40° F.) with high humidity to reduce shriveling.

OTHER STAPLES

Cheese: You can buy shredded cheese in bulk at warehouse clubs, such as Costco, and then put it into smaller freezer bags and freeze for up to four months. Brick cheese can be frozen but will not grate well when thawed. Check the use-by date and purchase cheese that has the longest shelf life.

White flour: White flour can be stored in a dark cupboard in a tightly sealed glass or plastic container for up to six months. A good way to store white flour for long-term storage is in #10 cans with an oxygen pack included. If you don't have the use of a dry pack canner, you can store flour in buckets, but be sure to freeze the flour first for a couple of days, take it out to thaw, and then freeze it again before putting it in the buckets with an oxygen pack. This will ensure that you don't open your bucket and find the flour full of weevils!

Whole grain wheat: Store wheat in buckets with tight-fitting lids. If you are not currently using whole wheat but want to, get your family used to it gradually. It's great in chocolate chip cookies, for example. Try using a combination of whole wheat and white flour in recipes. Grind only enough wheat for what you are making and store any leftover flour in the freezer.

To substitute whole wheat in your recipes, remember that wheat flour is heavier than white flour and needs more leavening. For example, when making my whole wheat bread (from hard white wheat), I use 2½ tablespoons of SAF yeast for four loaves. If you are using whole wheat (such as the soft white wheat) in a recipe using baking powder, increase the baking powder by 1 teaspoon for every 3 cups of whole wheat flour. Recipes using baking soda do not need to be adjusted. Soft white wheat has a storage life of eight years, so be sure you are using it and not just storing it and forgetting it.

Water: Storing water is also essential in case your water supply is interrupted for any reason. If you live in the country, as we do, a simple power outage will shut down the pump that brings water from your well. In this situation, you might want to have a generator or a water storage tank with gravity flow. In the city, any number of situations could contaminate the water supply, or it could be shut off in the event of certain natural disasters. You can live quite a while without food, but your body requires water for life, not to mention cleanliness. You can purchase bottled water or store water from your own source.

If you plan to store water in a barrel, be sure that it is clean. You can sanitize the barrel with a little bleach (no fragrance) and then rinse. Be sure that you put it in its permanent place *before* you fill it. Make sure you have a device that can be used to get the water out of the barrel. A simple pump is available at most stores that sell emergency essentials. Remember that water does not "go bad" like food does. If it develops an unpleasant smell or taste, using a filter or boiling it will resolve the problem. Over time, the oxygen goes out of the water and leaves it with a flat taste. Pouring the water back and forth between two containers will help improve the taste.

For extra water to wash with or flush the toilet, we fill up all our soda bottles with water and save them. You could also use large juice bottles. Because this water would not be for drinking, you don't have to worry about purification.

WHERE TO STORE IT NOW THAT YOU HAVE IT

Where to store food can be a challenge for some families, but you would be surprised how little space it takes for a three-month supply from which you will be using and replacing items. Organize your kitchen cupboards and pantry for maximum storage space. Check out the shelving accessories that allow you to store more by using the vertical space on a shelf. Narrow shelves, less than twelve inches deep, can often fit behind doors or in hallways.

The garage is not a good place to store food, unless it is well insulated. Food quality remains intact longer in cool temperatures, which is a consideration when storing for more than three months. In a garage, temperatures can fluctuate widely from season to season, depending on where you live.

If you live in a home with a basement, cold storage room, or root cellar, you only need shelves on which to store your food and water. Avoid storing food directly on cement floors. If you can afford it, there are several food storage shelf systems that rotate your cans for you. You load the cans from the back, and cans roll to the front as they are used.

Sometimes you need to get a little creative when it comes to finding a place to store extra food that won't fit in your kitchen pantry. Here are some creative places to store food:

Closets: Any kind of closet can become a good place for food storage, such as a closet under stairs. You can put in shelves, or just stack cases or buckets.

Under beds: If your beds offer enough space underneath, it's surprising how much can be stored there. You could put a bed up on cinder blocks to create more storage space, if needed.

Furniture: Jan had to get creative when they lived in their first home. She used #10 cans as the supports for bookcases. Buckets would work, too. Of course, she was using only long-term storage for this purpose. She also used cases of food to create the base for a round tabletop, over which she placed an attractive floor-length cloth.

Homemade root cellar: Jan decided she needed a root cellar, because she has no basement. However, she has a hillside behind her home in the country. She had the hill excavated to make room for an 8x10-foot steel container. It was then covered over and has become part of the hillside. It has a door for easy access and works great for her long-term food storage, as well as for storing fresh vegetables, such as potatoes. Jan may not have a basement, but she now has a root cellar. This is probably not feasible for most people; I only include it to demonstrate how thinking creatively about a problem can lead you to your own solutions.

RECIPES

In the recipes that follow, we will show the original ingredients and also substitutions that you can make from your food storage. For example, where milk is mentioned, we will show alternate ingredients, such as reconstituted powdered milk or canned evaporated milk that has been diluted by half. Eggs also have some alternatives, which we list with each recipe containing eggs. (For additional egg substitutions, see page 50.)

BREAKFAST RECIPES

If you want to cook without a mix, here are some delicious and simple recipes that you can make from storable items.

Oatmeal with Cinnamon and Raisins

⅓ cup raisins (or more, if desired)
3½ cups water
¼ teaspoon salt
2 cups old-fashioned oats
1 teaspoon vanilla
¼ teaspoon cinnamon

Combine raisins, water, and salt in a medium pan and bring to a boil over high heat. Stir in oats, vanilla, and cinnamon. Reduce heat to low, cover, and simmer for six minutes. Serve with a little brown sugar and milk. The raisins naturally sweeten the cereal, so you don't need as much sugar. Serves 4.

Granola

    2 cups rolled oats
    1½ cups rolled wheat
    ½ cup wheat germ
    ½ cup flaked coconut
    1 teaspoon salt
    5 tablespoons brown sugar

Mix above ingredients together.

    ¼ cup hot water
    5 tablespoons honey
    ¼ cup salad oil
    1 teaspoon vanilla

Preheat oven to 300° F.

Combine oats, wheat, wheat germ, coconut, salt, and brown sugar in a large bowl; set aside. In a separate bowl, combine water, honey, oil, and vanilla. Pour liquid mixture into dry mixture and combine well. Spread mixture on a rimmed baking sheet and bake 30 minutes. Remove from oven, stir with fork to separate, then let cool. Store in a covered, airtight container. Can be eaten by handfuls (messy, but good) or with milk for a delicious cold cereal.

Biscuits and Gravy

    1 pound lean breakfast sausage
    7 tablespoons flour
    3½ cups milk, divided (or use reconstituted powdered milk or 1¾ cups
        evaporated milk mixed with an equal amount of water)
    1 teaspoon salt
    ¼ teaspoon pepper
    1 recipe Buttermilk Biscuits or canned, refrigerated biscuits

In a large skillet, brown sausage over medium heat. Sprinkle enough flour onto meat to coat meat and soak up drippings. Pour in 3 cups of the milk. Add salt and pepper, adjusting to taste. Mix the remaining ½ cup milk with remaining flour to make a thin paste for thickening. When gravy comes to a boil, whisk in thickening paste and simmer until gravy is thickened sufficiently.

Prepare biscuits using the recipe below or use a biscuit mix or refrigerated biscuits. Serve gravy over top.

Buttermilk Biscuits

2 cups all-purpose flour
2½ teaspoons baking powder
⅛ teaspoon salt
⅓ cup shortening
¾ cup buttermilk (powdered cultured buttermilk works great)

Preheat oven to 450° F. Grease a cookie sheet and set aside.

In a medium bowl, combine flour, baking powder, and salt. Cut in the shortening until mixture is crumbly. Add buttermilk, stirring just until moistened. Turn dough onto a floured surface and knead for 1 minute. Roll out to ½-inch thickness. Cut with a biscuit cutter or drinking glass after dipping rim in flour. Place on prepared baking sheet and bake 10 to 12 minutes, or until golden brown.

German Pancakes

This recipe sometimes goes by the name Dutch Babies and can be baked in a soufflé dish instead of a 9x13-inch pan, if desired. If you use a soufflé dish, cut the pancakes in wedges to serve.

6 tablespoons butter
6 eggs (or use 6 frozen scrambled eggs, thawed, or reconstituted powdered eggs)
1 cup flour
¼ teaspoon salt
1 cup milk (or use reconstituted powdered milk or ½ cup evaporated milk and ½ cup water)

Preheat oven to 350° F.

Melt butter in a 9x13-inch baking dish; set aside. Add eggs, flour, salt, and milk to the jar of a blender and blend well. Pour egg mixture into melted butter and bake for 20 minutes. The pancake will puff up when ready to serve. Serve with lemon juice and powdered sugar or hot syrup. German pancakes are also delicious with canned or fresh peaches or fresh strawberries with hot syrup drizzled over the top.

Fruity Baked Oatmeal

As a delicious variation, replace peaches with ½ cup raisins (microwave in a little water for 1 minute to plump raisins). Stir in ¼ cup chopped nuts and 1 teaspoon vanilla before baking. This dish is good warm or cold.

3 cups rolled oats
½ cup brown sugar
2 teaspoons baking powder
1 teaspoon salt
1 teaspoon cinnamon
2 eggs, slightly beaten (or use 2 frozen scrambled eggs, thawed, or reconstituted powdered eggs)
1 cup milk (or use reconstituted powdered milk or ½ cup evaporated milk and ½ cup water)
½ cup butter, melted
1 apple, peeled and chopped, or rehydrated dried apples or freeze-dried apples
½ cup chopped fresh, frozen, or canned peaches
½ cup fresh, frozen, or canned blueberries

Preheat oven to 350° F.

Coat an 8-inch square baking dish with nonstick cooking spray; set aside.

In a medium bowl, combine the oats, brown sugar, baking powder, salt, and cinnamon. In a separate bowl, combine the eggs, milk, and butter; add to the dry ingredients. Stir in the apples, peaches, and blueberries. Pour into prepared pan and bake uncovered 40 to 45 minutes.

Whole Wheat Pancakes

- 3 cups whole wheat flour
- 3 tablespoons baking powder
- ½ teaspoon salt
- 3 eggs (or use reconstituted powdered whole eggs, frozen scrambled eggs, or even the Egg Substitute recipe below)
- 3 cups milk (or use reconstituted powdered milk or 1½ cups evaporated milk and 1½ cups water)
- ¼ cup oil or melted butter
- 1 teaspoon vanilla

Mix all ingredients well. Pour ⅓ cup batter for each pancake onto hot griddle. Turn when bubbles begin to form on top and edges begin to dry.

Egg Substitute

Use this recipe for each egg you'd like to replace.

- 1 teaspoon unflavored gelatin
- 2 tablespoons boiling water
- 2 tablespoons cold water

Whisk gelatin in boiling water to dissolve. Add cold water. Use in baked goods in place of eggs.

Blender Whole Wheat Pancakes

- 1½ cups milk (or use reconstituted powdered milk)
- 1 cup whole wheat grain
- 1 egg (or use reconstituted powdered whole eggs, frozen scrambled eggs, or Egg Substitute)
- 2 tablespoons butter or margarine
- 1 tablespoon honey
- ½ teaspoon salt
- 1 teaspoon baking powder

In the jar of a blender, mix milk and whole wheat kernels on high for about 3 minutes. Add egg, butter, honey, and salt. Mix for another 20 seconds or so. Add baking powder and pulse 3 times. Mixture should foam up and get very light. Cook immediately on hot oiled griddle.

Sourdough Pancakes

We love these pancakes, but you need to begin with a sourdough starter and must, therefore, plan ahead a bit unless you are using your sourdough starter regularly. There are two kinds of starters you can make: yeast and no yeast. Recipes for each are included after the pancake recipe.

1 cup sourdough starter
2 cups flour
2 cups milk (or use reconstituted powdered milk)
1 teaspoon salt
2 teaspoons baking soda
2 eggs
2 tablespoons sugar
3 tablespoons oil

The night before serving, add flour, milk, and salt to the 1 cup sourdough starter. Mix until smooth. Cover and let stand at room temperature overnight. (If doubling or tripling recipe, note that you can add as much as 6 cups flour, 6 cups milk and 3 teaspoons salt to the 1 cup starter.)

The next morning, before you start to make your pancakes, remove 1 cup of the mixture and place in refrigerator. This is to replace your start. (I've forgotten to complete this step more than once and had to make my starter all over again.)

To the remaining mixture add the baking soda, eggs, sugar, and oil. Mix well, then fry on hot griddle until bubbles form on top and edges begin to dry, then turn and complete cooking.

Yeast Sourdough Starter

1 package active dry yeast or 2 tablespoons SAF yeast
½ cup warm water
2 cups flour
2 cups lukewarm water

1 tablespoon sugar

1 teaspoon salt

In a glass or plastic mixing bowl (do *not* use a metal bowl), dissolve yeast in the ½ cup warm water. Stir in flour and the 2 cups lukewarm water, sugar, and salt. Beat until smooth. Put in glass container with lid, and let stand uncovered at room temperature for 3 to 5 days. Stir 2 to 3 times each day. Cover at night to prevent drying.

At the end of the 3 to 5 days, the starter should have a yeasty, not sour, smell. It is now ready to use in the sourdough recipes. Store in refrigerator between uses.

No-Yeast Sourdough Starter

1 cup milk (or use reconstituted powdered milk)

1 cup flour

Let milk sit in a glass jar at room temperature for 24 hours. Stir in the 1 cup flour and cover with a cloth. Set mixture in a warm place (80° F. is the ideal temperature) for 2 to 5 days, depending on how long it takes for it to bubble and sour.

To replace the portion of starter you use in recipes, add equal amounts of flour and milk to the remaining start and allow this to set out at room temperature for 24 hours. This should be done about every 14 days, at least. Store in refrigerator between uses.

Sourdough English Muffins

English muffins can be a great substitute for bread. Make sure you plan ahead when preparing this recipe.

1 cup sourdough starter

1 cup milk (or use reconstituted powdered milk)

3¾ cups flour, divided

1 tablespoon sugar

¾ teaspoon salt

½ teaspoon baking soda

3 tablespoons cornmeal

In a large mixing bowl, combine starter, milk, and 2 cups of the flour. Mix by hand, cover loosely, and set aside at room temperature for about 8 hours or overnight. The next morning, combine sugar, salt, baking soda, and ½ cup of the flour in a small bowl. Sprinkle over the dough and mix in.

Turn this stiff dough onto a board floured with remaining flour. Knead for 2 to 3 minutes, or until dough is no longer sticky. Add more flour if needed.

Roll dough out to a ¾-inch thickness. Use a 3-inch round cutter to cut out 12 muffins. Place muffins 1 inch apart on a cookie sheet that has been sprinkled with cornmeal. Cover with a cloth and set aside in a warm place to rise, about 45 minutes. Lightly grease and heat an electric griddle to 275° F. Transfer muffins to griddle and bake for about 8 to 10 minutes per side. Turn once. Serve warm from the griddle, or split and toast.

Homemade Maple Syrup

1 cup water
2 cups sugar
½ cup brown sugar
2 teaspoons maple flavoring
¼ cup corn syrup

In a medium saucepan, bring water to a boil over high heat. Add sugars and stir while simmering until completely dissolved. Add flavoring and corn syrup. Combine well before serving. Store any leftover syrup in the refrigerator.

Cinnamon Cream Syrup

1 cup sugar
½ cup corn syrup
¼ cup water
¾ teaspoon cinnamon
½ cup evaporated milk

In a small saucepan, combine sugar, corn syrup, water, and cinnamon. Bring to a boil, stirring constantly. Continue to stir while the syrup boils for

2 minutes. Stir in the evaporated milk. Remove from heat and cool 5 minutes before serving. Store any leftover syrup in the refrigerator.

Apple Spice Syrup

You can use cornstarch to thicken any favorite juice, such as cranberry or any citrus juice, to make a syrup. Just use the same method described in this recipe. Experiment with sugars and spices. This is a good substitute if you don't have corn syrup on hand.

¼ cup brown sugar
2 tablespoons cornstarch
¼ teaspoon ground allspice
⅛ teaspoon ground nutmeg
1¾ cups apple juice or cider

In a small saucepan, combine brown sugar, cornstarch, allspice, and nutmeg; mix well. Add juice or cider. Cook and stir over medium heat until syrup is bubbly and slightly thickened.

Mixer Whole Wheat Bread

This delicious bread takes only 8 ingredients. If you have a bread mixer, you'll have 4 fresh loaves of bread, hot from the oven, in about 90 minutes. And that includes grinding your wheat.

6 cups hot water
8 cups hard white wheat, ground
2½ tablespoons SAF instant yeast
2 tablespoons salt
½ cup oil
⅔ cup honey
⅓ cup vital gluten
3 tablespoons dough enhancer

Put the hot water into a large mixing bowl. Add 8 cups of the freshly ground whole wheat flour. (Flour will be quite warm.) Mix on low speed. While mixing, add yeast and salt. Stop mixer to add oil and honey. Start mixer again and add vital gluten. Add about 3 more cups of the flour.

Then add dough enhancer. By this time you'll need to turn your mixer to the higher speed. Add more flour, one cup at a time, until the dough begins to pull away from the sides of the bowl. When the sides of the bowl look fairly clean, continue mixing for 10 minutes.

Oil your hands (vinyl kitchen gloves work well) and the surface on which you will be turning out the dough. Form the dough into a large log. Cut into 4 equal parts. Knead each part into a loaf form and put in 4 separate greased bread pans.

Cover loaves lightly with plastic wrap and let rise until top is level with sides of the pan (this will take only about 25 minutes because the dough is so warm). Place pans in cold oven. Set oven to 350° F. and bake for 35 minutes. Remove from pans and brush tops with melted butter.

Baked Cinnamon Scones

2 cups whole wheat flour (soft white wheat flour works well in this type of recipe)
1 tablespoon baking powder
1 teaspoon baking soda
1 tablespoon sugar
½ teaspoon salt
½ teaspoon cinnamon
2 tablespoons oil
½ cup skim or buttermilk (or use reconstituted powdered milk or powdered buttermilk)

Preheat oven to 450° F.

Combine dry ingredients in a large bowl. Add oil and milk and stir with a fork until moistened and mixture clings to itself. Knead dough gently on a floured surface about 8 times. Divide dough into 3 parts. Roll each part ½-inch thick. Cut into 6 wedges or into 2-inch rounds. Place on nonstick baking sheets. Bake 10 to 15 minutes.

RECIPES FOR LUNCH AND DINNER

We are including the recipes for the meals in our sample lunch and dinner meal plans, plus several others for you to try. All recipes are family favorites and are easy to make from storable foods. Recipes show original ingredients and alternatives you can use from your food storage.

Quick Spaghetti with Meat Sauce

This sauce takes very little time to prepare and tastes like you simmered it for hours!

½ onion, diced, or ¼ cup dried minced onion flakes
1 pound ground beef
Seasoning salt
Pepper to taste
½ teaspoon onion powder
¼ teaspoon garlic powder
1 (26.5-ounce to 32-ounce) can or jar spaghetti sauce
1 (14.5-ounce) can diced Italian tomatoes
1 pound dry spaghetti noodles

Brown ground beef and diced onions in a large nonstick skillet over medium heat. (If you are using dried minced onion flakes, add them to the sauce.) Season meat with seasoning salt, pepper, onion powder, and garlic powder. Drain off any excess fat. Add spaghetti sauce and diced Italian tomatoes. Simmer, covered, while you cook the spaghetti noodles according to the package directions. Drain pasta and serve the sauce separately to be spooned over the noodles. Serves 6 to 8.

Macaroni and Cheese

6 tablespoons butter
⅓ cup flour
1 teaspoon salt
¼ teaspoon onion powder
2 cups milk (or use reconstituted powdered milk or 1 cup evaporated milk and 1 cup water)
8 ounces processed American cheese, such as Velveeta

2 cups dry elbow macaroni, boiled until just tender and drained)
Grated cheddar cheese (optional)
1 cup saltine or Ritz crackers, crushed (optional)
2 tablespoons butter, melted (optional

Melt the 6 tablespoons butter in a large saucepan over low heat. Whisk in flour, salt, and onion powder until smooth and bubbly. Slowly add milk while whisking. Simmer until mixture thickens. Cut the processed cheese into cubes and add to the white sauce. Stir until melted and smooth.

Pour sauce over macaroni and mix through. You can serve it this way, or make it into a casserole. To make a casserole, put the macaroni in a greased 9x13-inch casserole pan and pour sauce on top. Top with grated cheddar. Combine crushed crackers with the 2 tablespoons melted butter and put on top of the cheese. If you don't have cheddar cheese, just put the crushed, buttered crackers on top. Bake at 350° F. for about 25 minutes.

Easy Fettuccine Alfredo Primavera

1 (16-ounce) jar of your favorite Alfredo sauce, or 2 packets Alfredo
 sauce prepared according to directions on packet
2 cups cooked, chopped chicken
1 (12-ounce) package frozen California Mix vegetables
8 ounces fettuccine, prepared according to package directions and
 drained

Heat Alfredo sauce in a medium saucepan, or make up packets into sauce; add chicken pieces and vegetables and heat through. Serve over prepared fettuccine.

Homemade Ricotta Cheese

I make this to go into a baked ziti casserole. It is easy and delicious. It can be made with regular milk or with reconstituted non-instant powdered milk.

8 cups whole or 2 percent milk (or use reconstituted non-instant
 powdered milk)

½ teaspoon salt

3 tablespoons lemon juice or white vinegar

In a heavy-bottom 4-quart saucepan, heat milk and salt to boiling over medium-high heat. Stir to keep from scorching. When milk comes to a boil, stir in lemon juice or vinegar, cover, and remove from heat. Let stand 5 or 6 minutes.

With slotted spoon, stir gently as curds form and separate from the whey. Strain the curds through a mesh strainer or cheesecloth set in a colander. Rinse with cool water and use immediately or transfer to a clean bowl, cover, and refrigerate up to 5 days.

Baked Ziti

1 pound ziti

2 cups ricotta cheese

¼ cup grated Romano cheese

2 (8-ounce) cans tomato sauce

⅛ teaspoon black pepper, or to taste

2 cups grated mozzarella cheese

Preheat oven to 350° F.

Cook ziti according to package directions until al dente, stirring often. Drain, but do not rinse. Combine ricotta, Romano cheese, and tomato sauce (reserving ¼ cup of tomato sauce). Gently stir sauce into cooked ziti. Spread reserved ¼ cup tomato sauce in the bottom of a 9x13-inch pan. Add ziti mixture and top with mozzarella cheese. Cover loosely with foil and bake about 20 minutes, until cheese is thoroughly melted.

Beef-Barley Soup

When I make soup, I put in whatever vegetables I have on hand, regardless of what the recipe may call for. I call these soups my "very vegetable soups." Feel free to do the same here.

1 pound lean ground beef

1 onion, chopped, or ⅓ cup dried chopped onions

8 to 10 cups water

8 to 10 teaspoons beef bouillon
½ teaspoon onion powder
½ teaspoon garlic powder
¾ cup quick barley
4 or 5 carrots, sliced (can also use frozen or canned sliced carrots)
1 cup sliced fresh mushrooms (can also use canned mushrooms)
Salt and pepper to taste
1 cup medium shell pasta
Parmesan cheese (optional)

Brown ground beef with onion in a large pot over medium heat. Drain off any excess fat. Add water and bouillon and remaining ingredients, except for pasta. Cook 10 minutes. Add pasta and cook until tender. Serve with Parmesan cheese, if desired. Serves 6 to 8.

Chicken Corn Chowder

I often use Hormel Bacon Crumbles in soups and salads. The bacon is precooked and ready to toss into any recipe. It can be found in the same aisle as salad dressings. It is most economical when purchased in larger packages at warehouse stores, such as Costco.

4 medium potatoes, chopped, or canned potatoes, drained and cut up
½ onion, diced, or ¼ cup dried onion flakes
2 celery ribs, sliced, or ¼ teaspoon celery seed
1 (14-ounce) can chicken broth, or 2 cups water mixed with 3 teaspoons chicken bouillon
2 boneless, skinless chicken breasts or 1 (13-ounce) can chicken
2 cups milk (or use reconstituted powdered milk, or 1 cup evaporated milk and 1 cup water)
1 (15-ounce) can cream-style corn
1 (15-ounce) can whole kernel corn, drained
½ cup flour (to thicken)
½ teaspoon onion powder
¼ teaspoon seasoned salt
¼ teaspoon pepper
6 slices bacon, cooked and crumbled, or ⅓ cup Hormel Bacon Crumbles

Place potatoes, onions, and celery in a large pot. Pour chicken broth over the top. Broth should cover potatoes. Add more broth if needed. If using chicken breasts, cut into pieces and add. Bring to a simmer over medium-high heat and cook until potatoes are tender and chicken is no longer pink. When potatoes are done, stir in milk, cream-style corn, and whole kernel corn.

In a small bowl, mix flour and water to make a thin paste. Whisk this into the simmering soup and combine well. Let soup simmer until thickened. Add seasonings, adjusting to taste if needed. If using canned chicken, add it now. Just before serving, add crumbled bacon.

Quick Clam Chowder

2 cups cubed potatoes, fresh or canned
½ cup diced celery, or ½ teaspoon celery seed
½ cup diced onions, or ⅓ cup dried onion flakes
2 (6.5-ounce) cans minced clams, juice reserved
3 teaspoons chicken bouillon
¾ cup butter
¾ cup flour
4 cups milk (or reconstituted powdered milk)
Salt and pepper to taste
½ teaspoon onion powder

Place potatoes, celery, and onions in a large pot. Add clam juice and enough water to just cover the potatoes. Stir in chicken bouillon. If using canned potatoes, bring liquid to a boil and then simmer until onions are softened. If using fresh potatoes, bring to a simmer and cook until potatoes are tender. In a separate saucepan, melt butter and stir in flour to make a roux. Cook, stirring, over medium heat about 1 minute. Roux will take on a nutty, golden flavor and color. Gradually stir in milk, salt and pepper to taste, and onion powder. Add this mixture to the potatoes and simmer until thickened. Stir in clams and heat through.

Quick Salmon Chowder

1 (10.75-ounce) can condensed cream of potato soup
2 cups milk (or use reconstituted powdered milk)
1 (15-ounce) can cream-style corn
1 (14.75-ounce) large can salmon, drained, skin and any bones
    removed
1/8 teaspoon pepper, or to taste

In a large pan, combine soup and milk and stir until smooth. Stir in cream-style corn. Break salmon into chunks and add to soup. Add pepper, to taste. Stir over medium heat just until boiling.

Easy Taco Soup

1 pound ground beef
1/2 cup diced onion, or 1/4 cup dried onion flakes
1 package taco seasoning
2 (14.5-ounce) cans diced tomatoes
2 (15-ounce) cans whole kernel corn, drained
2 (15.5-ounce) cans chili beans or ranch-style beans, undrained
4 cups water
Crushed tortilla chips
Grated cheese, for topping (optional)

Brown hamburger and onions with taco seasoning in a large pot over medium heat. Add tomatoes, corn, beans, and water. Cover and simmer 15 minutes, or until heated through. Serve over crushed tortilla chips. Sprinkle with grated cheese.

Flour Tortillas

3 cups flour
2 teaspoons baking powder
1 teaspoon salt
3 tablespoons oil
1 cup warm water

Combine dry ingredients together in a large bowl. Add oil and warm water and combine with a spoon or your hands. Form dough into a large ball, then pinch off pieces of dough to make 12 balls. Use a floured rolling pin to roll out tortillas. Tortillas should be quite thin. Cook on a medium-high skillet (no oil) until warm through, flipping once. These make a great bread substitute and are wonderful just with a little butter. Use them in your favorite Mexican recipes.

Corn Tortillas

2 cups *masa harina* tortilla mix
½ cup warm water

Mix and fry according to package directions. Tortilla presses are inexpensive and make it a snap to make homemade tortillas.

Navajo Fry Bread

1 cup white flour
½ cup whole wheat flour
1 tablespoon sugar
½ teaspoon baking powder
¼ teaspoon salt
½ cup water
½ cup honey (optional)

Combine dry ingredients in a medium bowl. Add water and mix well. Knead dough on a floured board until it is smooth and elastic. Let dough rest 10 minutes, covered. Roll out dough until it is ½-inch thick. Cut into squares or circles. Heat oil in a deep fryer to 370° F. (If not using a deep fryer, add ½ inch of oil to large frying pan.) Maintain temperature carefully, adjusting heat as needed. Place dough pieces in hot oil, using tongs to turn as the pieces brown. Once golden brown, remove from oil with a slotted spoon and drain on paper towels. Serve drizzled with honey, or turn into a meal by topping with chili and grated cheese to make Navajo tacos.

Corn Fritters

1 cup flour
2 teaspoons baking powder
¼ teaspoon salt
2 eggs, beaten (or frozen scrambled eggs, thawed, or Egg Beaters)
1 cup milk (or reconstituted powdered milk)
1 cup whole kernel corn, drained
Oil
Powdered sugar, for dusting

In a medium bowl, combine flour, baking powder, and salt. Add eggs and milk and stir together. Add corn and mix. Cover a skillet with about ¼ inch oil. Heat oil, then drop batter by large serving spoonfuls onto hot oil. Cook about 2 minutes on each side until golden brown and done all the way through. Drain on paper towels and sprinkle with powdered sugar. Serve as a snack or side dish with dinner.

Sunshine Fruit Salad

1 small package vanilla pudding (cook and serve)
1 small package tapioca pudding (cook and serve)
1 (20-ounce) can pineapple chunks, juice drained and reserved
2 (11-ounce) cans mandarin oranges, juice drained and reserved
1 (15-ounce) can fruit cocktail, juice drained and reserved
3 tablespoons frozen orange juice concentrate
1 or 2 bananas (if available)

Put both pudding mixes in a large saucepan. Combine reserved juices, orange juice concentrate, and enough water to equal 3 cups. Stir liquid into pudding mixes and cook over medium heat, stirring constantly, until pudding comes to a boil and thickens. Remove from heat and chill. (I usually do this the night before, or in the morning.) Once pudding is chilled, fold in well-drained fruit and any fresh fruit you have on hand, such as bananas or grapes.

Cowboy Beans

1 pound ground beef
½ cup diced onion (optional)
½ cup diced bell pepper (optional)
2 (15-ounce) cans pork and beans
1 teaspoon mustard
¼ cup ketchup
½ cup brown sugar

Brown ground beef with onion and bell pepper in a large pan over medium heat. Drain off any excess fat. Stir in pork and beans and remaining ingredients. Simmer for 5 minutes, until sugar has dissolved.

Quick Spanish Rice

Serve this as a side dish, or add chicken and make it a meal.

3 cups water
2 tablespoons butter
1½ cups rice
3 teaspoons tomato bouillon (Knorr brand, usually found in Latino section of store)
1 to 2 cups small chicken pieces (optional)

Bring water and butter to a boil in a large pan. Add rice, bouillon, and chicken pieces if using. Reduce heat, cover, and simmer on low for 20 minutes without lifting lid.

Sweet and Sour Green Beans

3 slices bacon, chopped, or 2 tablespoons Hormel Bacon Crumbles
1 tablespoon olive oil (optional)
½ onion, diced, or ¼ cup dried chopped onion
1 (16-ounce) can cut green beans (drain and reserve ½ cup liquid)
2 teaspoons cornstarch
¼ teaspoon dry mustard
1 tablespoon brown sugar
1 tablespoon vinegar

If using fresh bacon, cook in a hot skillet until crisp. If using Hormel Bacon Crumbles, sauté in 1 tablespoon olive oil. Remove bacon from pan, leaving some of the grease. (If using fresh onions, add and cook until tender.) Stir the ½ cup reserved liquid, cornstarch, and dry mustard into the onions and cook over medium-high heat until thickened. Blend in sugar and vinegar. Add beans and heat. Top with bacon before serving.

Surprise Baked Potatoes

¼ cup butter or margarine, melted
⅓ cup grated Parmesan or Romano cheese
5 or 6 small potatoes, scrubbed and cut in half lengthwise

Preheat oven to 400° F.

Pour melted butter on a large, rimmed cookie sheet covered with foil. Spread evenly. Sprinkle the cheese over the butter. Place the potatoes, cut side down, on top of the cheese. Bake about 40 minutes, or until tender when pierced with a fork.

Cheesy Potatoes with Ham

2 (10.75-ounce) cans cream of chicken soup
2 cups sour cream (or use powdered sour cream, reconstituted according to directions)
1½ cups shredded cheddar cheese
½ teaspoon onion powder
2 cups diced ham
12 cooked potatoes, grated, or 6 cups frozen or reconstituted hash brown potatoes
Salt and pepper to taste
½ cup butter or margarine, melted
2 cups crushed crackers or cornflakes

Preheat oven to 350° F.

Combine soup, sour cream, cheese, onion powder, and diced ham in a large bowl. Stir in potatoes and salt and pepper to taste. Spread mixture in a 9x13-inch pan. In a separate bowl, combine melted butter and crushed crackers or cornflakes. Sprinkle on top of potatoes and bake about 45 minutes.

Chicken Enchilada Casserole

1 tablespoon oil
1 or 2 (13-ounce) cans chicken
1 medium onion, finely chopped, or ½ cup dried onion flakes
1 (15.5-ounce) can olives, sliced
1 (10.75-ounce) can cream of chicken soup
1 (10.75-ounce) can cream of mushroom soup
1 cup sour cream (optional)
1 (10.5-ounce) can green enchilada sauce
1 (4-ounce) can diced green chilies
20 corn tortillas
3 cups shredded cheddar cheese (or whatever cheese you have on hand)

Preheat oven to 350° F.

Grease the bottom of a 9x13-inch casserole pan with the oil and set aside.

Mix chicken, onions, and olives in a small bowl. In a medium pan, combine soups, sour cream, green enchilada sauce, and chilies. Stir in chicken mixture and heat to a simmer. Layer corn tortillas so they overlap in the bottom of the prepared pan. Layer half of the sauce and chicken mixture on top. Sprinkle with half of the cheese. Repeat layers and top with cheese. Bake 30 minutes or until bubbly. Serves 6 to 8.

Chicken or Beef Gravy over Noodles

You can make this meal using the basic white sauce in the recipe and flavor it with either chicken or beef bouillon as written. Or, you can simply use 2 or 3 gravy packets, prepared according to the directions on the package. Omit the first 6 ingredients if using a packet. Add your favorite canned or fresh cooked meat and serve over noodles or mashed potatoes.

3 to 4 tablespoons butter
3 to 4 tablespoons flour

2 cups milk (or reconstituted powdered milk)
2 teaspoons chicken or beef bouillon
⅛ teaspoon pepper
Dash onion powder (optional)
1 pound cooked meat, such as chicken or beef, or 2 (13-ounce) cans
    chicken or beef
1 pound cooked noodles or 1 recipe mashed potatoes or rice

Melt butter in a large saucepan over medium-low heat. Whisk in flour and stir until bubbly. Slowly add milk while whisking. Add bouillon and seasonings. Simmer until thickened. Add meat of your choice and heat through. Serve over noodles, mashed potatoes, or rice.

Baked Chicken Curry

You can substitute 3 cans of chicken for the chicken breasts in this dish for a very storage-friendly meal. Serve over rice or noodles.

3 or 4 chicken breasts, cut in half, or 3 (13-ounce) cans chicken
1 (10.75-ounce) can cream of chicken soup
½ cup mayonnaise
1 teaspoon lemon juice
½ teaspoon curry powder
2 cups shredded cheese

Preheat oven to 350° F.

Arrange chicken pieces in a baking dish. Mix soup, mayonnaise, lemon juice, curry powder, and cheese together and pour over the chicken. Bake 30 minutes or until chicken tests done (165° F. on meat thermometer, or until juices run clear). Serve over rice or noodles.

Chicken and Rice

You can substitute 2 cans of chicken for the chicken breasts in this dish for a delicious storage-friendly meal.

1 (10.75-ounce) can cream of mushroom soup
1 (10.75-ounce) can cream of chicken soup
1 soup can filled with water
1 package onion soup mix

1½ cups rice
2 chicken breasts, cut into pieces

Preheat oven to 350° F.

Mix soups, water, onion soup mix, and uncooked rice together. Spread half of the soup mixture on the bottom of a lightly greased 9x13-inch pan. Arrange canned or fresh chicken on top. Cover with remaining soup mixture. Bake, covered, 1 hour.

Poppyseed Chicken

4 chicken breast halves, cooked and torn into pieces
1 (27-ounce) can cream of chicken soup
1 (16-ounce) tub sour cream
1 (4-ounce) can diced green chilies
1 tube Ritz crackers
½ cup butter or margarine, melted
2 tablespoons poppy seeds

Preheat oven to 350° F.

Combine chicken, soup, sour cream, and green chilies. Spread in a 9x13-inch baking dish. Crush Ritz crackers and sprinkle evenly over top. Drizzle melted butter over that. Sprinkle with poppy seeds. Bake 30 to 40 minutes or until bubbly and topping is lightly browned.

Chicken and Stuffing Casserole

1 (6-ounce) package stuffing mix, prepared according to package
   directions
1 (13-ounce) can chicken, or leftover chicken or turkey pieces
1 (10.75-ounce) can cream of chicken soup
1 cup milk (or use reconstituted powdered milk, or ½ cup evaporated
   canned milk mixed with ½ cup water)

Preheat oven to 350° F.

Spread prepared stuffing evenly in an 8x8-inch pan. (Double recipe for 9x13-inch pan.) Drain canned chicken and crumble over stuffing.

Mix soup and milk together and stir until smooth. Pour evenly over stuffing and chicken. Bake 30 minutes.

Chicken Puffs

¼ cup butter, softened
1 (8-ounce) package cream cheese
½ teaspoon onion powder
2 cups diced, cooked chicken, or 2 cans chicken
2 packages refrigerated crescent roll dough
1 cup butter or margarine, melted
1 cup bread or cracker crumbs
1 packet chicken gravy mix, prepared according to package
    directions

Preheat oven to 375° F. Grease a large baking sheet and set aside.

Cream butter or margarine together with cream cheese until smooth. Add onion powder and mix well. Fold in chicken pieces. Roll out a triangle of crescent dough and put 1 tablespoon of the chicken mixture on top. Roll up the dough and seal edges. Roll each puff in melted butter and then in bread or cracker crumbs. Repeat with each crescent dough triangle. Place on prepared baking sheet and bake 20 to 30 minutes. Make chicken gravy according to package directions. Serve puff with a spoonful of gravy over the top.

Chicken with Beans

½ onion, chopped, or ¼ cup dry chopped onions
2 stalks celery, chopped, or ¼ teaspoon celery seed
1 tablespoon oil
½ cup ketchup
½ cup brown sugar
1 teaspoon prepared mustard
1 tablespoon white vinegar
1 teaspoon salt
2 cups cooked chicken, chopped into pieces
1 (16-ounce) can red kidney beans
1 (16-ounce) can baked beans

Preheat oven to 350 F.

Sauté onion and celery in oil over medium heat until tender. In a separate bowl, mix ketchup, brown sugar, mustard, vinegar, and salt together; set aside. (If using dry chopped onions and celery seed, add to the ketchup mixture and skip the first step). Place chicken and beans in skillet and stir with onion and celery. Pour ketchup mixture over chicken and beans. Heat to warm through. Spoon into a medium casserole dish and bake 30 minutes or until bubbly and hot.

Sweet and Sour Chicken

1 (20-ounce) can pineapple chunks, juice reserved
2 cups sugar
1 cup vinegar
6 tablespoons soy sauce
8 tablespoons ketchup
½ cup flour
1 bell pepper, chopped, or ½ cup frozen chopped bell pepper
1 onion, cut into 1-inch chunks, or ¼ cup dried chopped onion
1 pound chicken breasts
1 tablespoon olive oil

Measure reserved pineapple juice and add enough water to get 1 cup total liquid. Pour into saucepan, along with sugar, vinegar, soy sauce, ketchup, and flour. Mix well and bring to a boil. Boil for 1 minute, stirring constantly. Add vegetables and heat through.

Cut chicken into chunks and sauté in 1 tablespoon olive oil until cooked through. Mix with sauce and serve over rice.

Sweet and Sour Pork

The sauce portion of this recipe can be made alone, poured over browned country-style ribs, and baked 1 hour at 350 degrees F. or until tender, for a great variation.

1 pound pork chunks
½ cup soy sauce
½ cup red wine vinegar

Tempura batter, prepared according to package (or use recipe
    below)
1 cup sugar
½ cup vinegar
½ cup water
3 tablespoons soy sauce
4 tablespoons ketchup
¼ cup flour

Marinate pork chunks for 15 minutes in ½ cup soy sauce and ½ cup red wine vinegar. Prepare tempura batter, then dredge pork in batter and deep fry until done, turning often. (To deep fry, melt shortening in a deep pan and heat to 350 degrees F. Maintain temperature carefully, adjusting heat as needed. Remove fried pieces from oil with a slotted spoon and drain on paper towels.) It takes about 5 minutes for small pieces to cook.

Prepare sauce by combining remaining ingredients in a saucepan. Bring to and maintain a boil, stirring constantly, for 1 minute. Stir in pork chunks. Serve pork and sauce over rice.

Tempura Batter

1 cup flour
1 cup cold water
1 egg (or use a thawed frozen scrambled egg)

Measure flour in a bowl and add the cold water. Stir slightly until flour is lumpy. (Do not mix too well.) Add a cold beaten egg. Stir slightly, but do not mix well. Batter must be cold at all times. Too much mixing or letting the batter turn warm will make it doughy.

Shepherd's Pie

1 pound lean ground beef
Salt and pepper to taste
1 (10.75-ounce) can tomato soup
1 can cut green beans

2 cups mashed potatoes (or use equal amount of reconstituted
dehydrated potatoes)
½ cup grated cheese

Preheat oven to 350° F.

Brown ground beef in a large skillet over medium-high heat, seasoning with salt and pepper to taste. Drain off any excess fat. Stir in soup and green beans. Place in a 2-quart casserole dish. Top with mashed potatoes and cheese. Bake 30 minutes.

Tamale Pie

1 pound lean ground beef
1 tablespoon chili powder
1 small onion, diced, or ¼ cup dry onion flakes
1 teaspoon cumin
1 cup medium salsa
1 (14.75-ounce) can whole kernel corn, drained
1 (15.5-ounce) can pitted olives
4 cups water
1 cup cornmeal
1 teaspoon salt
½ cup grated cheddar cheese

Preheat oven to 350° F.

Brown ground beef and onion in a large skillet over medium heat. Drain off any excess fat. Stir in chili powder and cumin; cook a few minutes more. Remove from heat and stir in salsa, corn, and olives.

In medium saucepan, heat water to boiling; whisk in cornmeal and salt. Cook over medium heat for 5 minutes, stirring frequently until thickened like a batter. Pour half of the cornmeal mixture into a shallow 2-quart casserole. Spoon beef mixture over cornmeal; spoon remaining cornmeal over beef. Sprinkle grated cheese on top. Bake 45 minutes. Let stand about 15 minutes before serving.

Mexican Meat Mix

Make this meat mix ahead of time and freeze. Then you can use it in any Mexican dish: tacos, enchiladas, tamales, etc.

1 (5-pound) beef or pork roast
3 tablespoons vegetable shortening
1 (4-ounce) can chopped green chilies
1 (7-ounce) can green chili salsa
1 teaspoon garlic powder
4 tablespoons flour
4 teaspoons salt
1 large onion, diced, or ½ cup dry onion flakes
1 teaspoon ground cumin

Wrap meat in foil and bake at 200° F. for twelve hours. You can also use a slow cooker, if desired. Carefully unwrap meat and drain, save juices. Cool meat, remove any bones, and shred. In a large skillet, melt shortening and add onion and green chilies. Sauté briefly and add remaining ingredients. Cook 1 minute over medium-low heat. Stir in reserved meat juices and shredded meat. Cook 5 minutes more until thick. Cool. Put 3 cups each into two 1-quart freezer bags or containers. Use within 6 months. Recipe can be doubled.

Mexican Lasagna

1½ to 2 pounds lean ground beef
1 (10.75-ounce) can cream of mushroom soup
1 (10.75-ounce) can cream of chicken soup
1 (15-ounce) can enchilada sauce
1½ cups salsa (mild, medium, or hot, according to your taste)
2 cups grated cheddar or Monterey Jack cheese
20 flour tortillas (homemade tortillas work great)

Preheat oven to 350° F.
Grease a 9x13-inch casserole dish and set aside.
Brown ground beef in a large skillet over medium heat. Drain off excess fat. Mix beef with soups, enchilada sauce, and salsa. Heat until warm. Place a layer of tortillas, topped with a layer of meat sauce, and a thin

layer of cheese in prepared pan. Repeat layers, spreading remaining cheese on top. Bake 30 minutes, until cheese is melted and sauce bubbles.

Corned Beef Casserole

1 (10.75-ounce) can cream of mushroom soup
1 (10.75-ounce) can cream of chicken soup
¼ cup dried minced onion
2 cups milk (or use reconstituted powdered milk or 1 cup evaporated milk and 1 cup water)
1 (12-ounce) can corned beef
2 cups cooked noodles
2 cups grated cheddar cheese
1 cup crushed cornflakes or crackers
¼ cup butter, melted

Preheat oven to 350° F. Grease a 9x13-inch pan and set aside.

Mix soups, onion, and milk together until smooth. Dice corned beef and fold into soup mixture. Spread noodles in the bottom of the prepared pan. Pour soup mixture over top and lightly stir with noodles. Top with grated cheese. Mix crushed cornflakes or crackers with melted butter and sprinkle over the top. Bake 30 minutes, or until cheese is melted and sauce is bubbly.

Rice and Black Beans with Beef

1 pound lean ground beef
½ onion, chopped, or ¼ cup dry chopped onion
¼ teaspoon onion powder
⅛ teaspoon garlic powder
1 (8-ounce) box seasoned rice and black beans (boxed dinner)
1 (14.5-ounce) can diced tomatoes
1 (4-ounce) can diced green chilies
1 (12-ounce) package frozen corn, or 1 (15-ounce) can whole kernel corn, drained
1½ cups water
¼ cup chopped cilantro (optional)
1 cup shredded cheddar cheese

Cook beef in a large skillet over medium heat with onion and seasonings, breaking beef into small pieces while it cooks. Stir in rice and beans, tomatoes, green chilies, corn, and water.

Bring to a boil, then reduce heat to simmer. Cover and cook 20 minutes until water has been absorbed. Serve with chopped cilantro and cheese sprinkled on top.

Easy Creamed Tuna

4 tablespoons butter or margarine
4 tablespoons flour
½ teaspoon salt
¼ teaspoon pepper
¼ teaspoon onion powder
2 cups milk (or reconstituted powdered milk)
2 or 3 (5-ounce) cans tuna (depending on size of your family)
½ cup frozen peas (optional)

Melt butter in a medium saucepan over medium heat. Add flour, salt, pepper, and onion powder, stirring until smooth and bubbly. Add milk and heat to boiling, stirring constantly for 1 minute to thicken. Add tuna and peas and heat through. Serve over toast, rice, or noodles.

Salmon Loaf

1 (14.75-ounce) can salmon
1½ cups crushed saltine crackers
1 egg, slightly beaten, or frozen scrambled eggs, thawed, or
    reconstituted powdered eggs
½ cup diced green pepper
½ cup diced onion, or ¼ cup dry chopped onions, reconstituted
¼ cup milk (or use reconstituted powdered milk or ¼ cup evaporated
    milk)
½ teaspoon Worcestershire sauce
Dash hot pepper sauce (optional)
⅛ teaspoon pepper

Preheat oven to 350° F.

In a medium bowl, stir together salmon, crackers, egg, green pepper, and onion. Mix in milk, Worcestershire sauce, and hot pepper sauce, if using. Season with black pepper. Mix well with your hands and spread in 9x9-inch pan. Bake for about 45 minutes, until top is golden brown.

Tuna-Potato Patties

½ cup mashed potatoes, fresh or rehydrated
1 (5-ounce) can water-packed tuna, drained and crumbled
¼ cup seasoned bread crumbs
¼ cup celery, finely chopped, or ¼ teaspoon celery seed
2 to 3 drops hot pepper sauce (optional)
Tartar sauce (optional)

Mix all ingredients together and shape into patties about ½-inch thick. In a hot skillet, fry over medium heat for about 4 minutes on each side. Serve with tartar sauce, if desired.

Salmon Patties

1 (7-ounce) can salmon
½ cup cracker crumbs
2 eggs, beaten (or use 2 frozen scrambled eggs, thawed, or
    reconstituted powdered eggs)
2 tablespoons lemon juice
1 teaspoon parsley flakes
¼ teaspoon dill weed
⅛ teaspoon onion powder
⅛ teaspoon pepper, or to taste

Drain salmon and remove any skin. Combine salmon, cracker crumbs, eggs, and lemon juice. Mix in seasonings. Shape into patties. Spray a skillet with cooking spray, or use a small amount of olive oil. Fry patties on both sides until golden brown. Makes about 4 patties.

Quick Chili

1 pound lean hamburger
Salt and pepper to taste
1 (15-ounce) can diced tomatoes
1 (15-ounce) can red kidney beans, drained
2 (15-ounce) cans red beans, drained
1 (8-ounce) can tomato sauce
½ onion, diced, or ¼ cup dried onion flakes
1 (4-ounce) can diced green chilies, with liquid
½ cup diced celery (optional)
1 tablespoon chili powder
½ teaspoon ground cumin
¼ teaspoon ground black pepper
¼ teaspoon garlic powder

In a large skillet, brown hamburger over medium heat, breaking into small pieces with spatula. Add a dash of salt and pepper while cooking. Drain fat from meat. Transfer meat into large pan and add the remaining ingredients, stirring to combine. Simmer over medium low heat for 50 to 60 minutes, stirring occasionally.

DESSERT AND TREAT RECIPES

We've included a variety of dessert and treat recipes for snacks, family nights, and so on. Each recipe includes the original ingredients and suggestions for alternative ingredients that can be used from your food storage.

Whole Wheat Chocolate Chip Cookies

5½ cups whole wheat flour (or 3 cups whole wheat flour and 4 cups rolled oats)
2 cups butter, softened
1¾ cups brown sugar
1½ cups white sugar
3 eggs (or use frozen scrambled eggs, thawed, or reconstituted powdered eggs, or equivalent Egg Substitute—see page 62)

1 tablespoon vanilla

1 teaspoon salt

1½ teaspoons baking soda

3 cups total any combination of chocolate chips, raisins, M&M's, butterscotch chips, or nuts

Preheat oven to 375° F.

Cream butter, brown sugar, white sugar, eggs, and vanilla in a large bowl. Mix in flour, salt, and soda. (If using whole wheat flour and rolled oats, blend to a flour in the blender.) Add your mix-ins.

Drop by spoonfuls onto greased cookie sheets and bake for about 7 minutes. Remove when cookies are barely brown and the tops are no longer shiny.

Easy Chocolate Chip Pumpkin Cookies

Please note that these cookies do not brown and may look undercooked. They are done when just firm. The end result is very moist and delicious as well as nutritious.

1 box spice cake mix

1 (15-ounce) can pumpkin

1 bag chocolate chips

Preheat oven to 375° F.

Combine ingredients in a large bowl and mix well. Drop by teaspoonfuls onto greased cookie sheet. Bake 10 to 12 minutes.

Peanut Butter Oatmeal Cookies

½ cup crunchy peanut butter

½ cup brown sugar

¼ cup butter, softened

1 cup flour

½ teaspoon baking soda

1½ teaspoons baking powder

1 egg, beaten (or use 1 frozen scrambled egg, thawed, or Egg Substitute—see page 62)

⅔ cup old-fashioned oats

½ cup chopped peanuts (optional)
2 tablespoons milk

Preheat oven to 375° F.

In a large bowl, beat the peanut butter, brown sugar, and butter until creamy. Mix dry ingredients and add to batter. Add egg, oats, peanuts, and milk. Stir into a soft dough. Roll a piece of dough into a walnut-sized ball. Place dough balls on greased baking sheets, spacing evenly. Slightly flatten each ball with the bottom of a glass. Bake about 9 to 12 minutes, or until lightly browned. Cool on the baking sheet for 2 minutes before transferring to a wire rack to finish cooling.

Shortbread Cookies

For Brown Sugar Shortbread Cookies, replace white sugar with ¾ cup brown sugar and shape dough into balls to bake. Increase temperature to 325° F. and bake 10 to 12 minutes.

1 cup butter, softened
½ cup sugar
2½ cups flour

Cream butter and sugar until light and fluffy. Add flour and mix well. Chill several hours. Divide in half. Roll out dough ¼- to ½-inch thick on ungreased cookie sheet. With a sharp knife, score dough into squares or triangles. Bake at 300° F. for 30 minutes, or until light brown. Cut cookies using the score lines as a guide. Remove to cooling rack.

Chocolate Sandwich Cookies

2 Devils Food cake mixes
1 cup shortening, margarine, or butter
4 eggs (or use frozen scrambled eggs, thawed, or reconstituted
powdered eggs, or Egg Substitute—see page 62)
1 recipe of your favorite frosting

Preheat oven to 350° F.

Combine all ingredients in a large bowl and mix well. Batter will be thick. Scoop equal sized balls onto greased cookie sheets and bake about

9 minutes. Cool on rack. Frost half of the cookies and top with remaining half of the cookies.

No-Bake Honey Bites

½ cup butter

2 tablespoons milk (or use evaporated milk or reconstituted powdered milk)

1 cup flour

¾ cup honey

¼ teaspoon salt

1 teaspoon vanilla

1½ cups flaked coconut

2 cups crisp rice cereal

Melt butter in a large saucepan over low heat. Blend in milk, flour, honey, and salt; mix thoroughly. Cook over medium heat, stirring constantly until dough leaves side of pan and forms a ball. Remove from heat. Stir in vanilla and 1 cup of the coconut. Cool. Add cereal. Shape into 1-inch balls. Roll in remaining coconut. Store in refrigerator. Makes about 42 candy bites.

No-Bake Crispy Peanut Butter Treats

¼ cup butter or margarine

1 (12-ounce) bag marshmallows

¼ cup light corn syrup

½ cup creamy peanut butter

⅓ cup milk chocolate chips

4½ cups crisp rice cereal

Line a cookie sheet with waxed paper. Melt butter in a large saucepan over low heat. Add marshmallows, stirring constantly until melted. Remove from heat and add corn syrup. Blend well, then add peanut butter and chocolate chips. Stir until melted and all is well blended. Add cereal and stir to coat evenly. Cool slightly. Shape into 1½-inch balls and place on prepared cookie sheet. Makes approximately 4 dozen.

No-Bake Cereal Bars

2 cups corn syrup
1 cup sugar
1 (40-ounce) jar chunky peanut butter
6 cups toasted oat cereal, such as Cheerios
6 cups crisp rice cereal

In a large saucepan, cook corn syrup and sugar, stirring constantly, until sugar is dissolved. Remove from heat and add peanut butter, mixing well. Stir in both cereals and quickly spread into two lightly greased 10x15-inch pans. Cut into bars while still a little warm.

Caramel Popcorn

1 cup butter
1 cup brown sugar
½ cup white corn syrup
1 gallon popped popcorn

Combine butter, brown sugar, and corn syrup in a heavy pan and bring to a rolling boil, then boil for 2 minutes longer. Pour over 1 gallon of popped popcorn and mix. Shape into popcorn balls, if desired.

Easy Hot Fudge Sauce

1 bag chocolate chips
½ cup butter
2 cups powdered sugar
1 (12-ounce) can evaporated milk
1 teaspoon vanilla

Melt chocolate chips and butter in a heavy saucepan over medium-low heat. Stir in sugar and evaporated milk. Bring to a gentle boil and boil for 5 minutes, or until thickened. Remove from stove and add vanilla. Sauce will thicken more as it cools. Makes about 1 quart.

Homemade Sweetened Condensed Milk

1 cup powdered milk
1/3 cup boiling water
2/3 cup sugar
3 tablespoons butter

Combine all ingredients in the jar of a blender and mix on slow speed for a minute or so. Increase speed and mix until smooth and sugar is dissolved. This recipe is equal to 1 store-bought can sweetened condensed milk.

Raisin Nut Pudding

1 tablespoon butter, softened
1/2 cup sugar
1/2 cup milk (or reconstituted powdered milk or 1/4 cup evaporated
    milk and 1/4 cup water)
1 cup flour
1/2 teaspoon baking soda
1/2 teaspoon nutmeg
Pinch salt
1 teaspoon vanilla
1/2 cup raisins
1/4 cup chopped nuts
1 recipe Quick Caramel Sauce (see below)

Preheat oven to 375° F.

Cream butter and sugar together in a large bowl. Stir in milk, then dry ingredients. Transfer to a large baking dish.

Prepare Quick Caramel Sauce and pour over raisin mixture.

Bake 30 minutes. Cool slightly. Serve warm with ice cream or whipped topping, if you have it. Otherwise, it's delicious all by itself.

Quick Caramel Sauce

2 cups water
1 1/2 cups brown sugar
2 tablespoons butter

Bring water to a boil in a medium saucepan, add sugar and butter, then heat, stirring until sugar is completely dissolved.

World War II Victory Cake

1 cup raisins
1 cup brown sugar
1 cup water
½ cup shortening, margarine, or butter
1 teaspoon cinnamon
1 teaspoon nutmeg
1 teaspoon ground cloves
2 cups flour
1 teaspoon baking soda
½ teaspoon salt
½ cup walnuts, chopped

Preheat oven to 300° F. Grease a 9x9-inch pan.

Combine raisins, sugar, water, shortening, and spices in a large saucepan. Heat to boiling and simmer 2 minutes. Cool. Combine flour with baking soda and salt and stir into raisin mixture. Add walnuts. Pour batter into prepared pan and bake 50 to 60 minutes, just until cake tests done.

No-Egg Chocolate Cake

You can mix this cake right in the pan.

1⅓ cups flour
2¼ cups sugar
1 cup unsweetened cocoa powder
2 teaspoons baking soda
1 teaspoon salt
¾ cup oil
2 tablespoons distilled white vinegar
1 tablespoon vanilla
2 cups warm water
1 bag chocolate chips (semi-sweet or milk chocolate)

Put flour, sugar, cocoa, baking soda, and salt in an ungreased 9x13-inch pan. Stir with a whisk until blended. Make 3 wells in mixture with a wooden spoon. Pour oil into one well, vinegar into another, and vanilla into the third. Pour 2 cups lukewarm water over all. Stir until blended, scraping sides and corners. Use a paper towel to wipe batter off top edge of pan.

Bake 30 minutes, or until done. Remove to a wire rack and immediately sprinkle with chocolate chips. Let stand until chips are soft and then spread melted chips over the cake. Cool before cutting.

No-Fail Pie Crust

Want to make a single-crust pie without the hassle of rolling it out? Try this delicious pie crust that turns out tender and flaky every time.

1 cup flour
¼ cup powdered sugar
¼ teaspoon salt
¼ cup nuts, chopped fine (optional)
½ cup butter or margarine, melted

Preheat oven to 400° F.

Put dry ingredients into 9-inch pie plate. Pour melted butter or margarine over the top and mix well with fork or whisk until all is moistened. Press mixture evenly into pie plate. Flute edges and bake about 8 minutes or until lightly browned.

Buttercream Frosting

This is a multi-purpose frosting that can be colored with food coloring, flavored with extracts, and/or doubled as needed.

½ cup butter, softened
4 cups powdered sugar
⅓ cup milk (approximately, may need a little more)
1 teaspoon vanilla

Cream softened butter and powdered sugar. Add milk and vanilla and beat on high for 2 to 3 minutes, or until light and fluffy.

Chocolate Fudge Frosting

½ cup butter
⅔ cup unsweetened cocoa powder
4 cups powdered sugar
⅓ cup milk (approximately, may need a little more)
2 teaspoons vanilla

Melt butter in saucepan with cocoa powder over low heat or in a double boiler, stirring often. Remove from heat and add powdered sugar, milk, and vanilla. Beat until smooth and creamy. Use immediately. This is a warm frosting that must be spread before it cools. Great for frosting cupcakes or Bundt cakes.

Brownies in a Pan

¾ cup unsweetened cocoa powder
⅔ cup plus ¼ cup shortening
2 cups sugar
4 eggs (or use frozen scrambled eggs, thawed, or reconstituted
     powdered eggs, or Egg Substitute—see page 62)
1 teaspoon vanilla
1¼ cups flour
1 teaspoon baking powder
1 teaspoon salt

Preheat oven to 350° F.

Combine cocoa and shortening in a large saucepan and melt together over medium heat. Remove from heat and stir in sugar, eggs, and vanilla until well blended. Add flour, baking powder, and salt.

Pour into a greased 9x13-inch pan and bake 23 to 25 minutes, or until brownies start to pull away from the sides. Do not overbake. Cool and cut.

No-Bake Cheesecake

1 (8-ounce) package cream cheese, softened
1 (14-ounce) can sweetened condensed milk
⅓ cup lemon juice

1 teaspoon vanilla

1 graham cracker pie crust

Beat cream cheese until smooth and creamy. Add milk and mix well. Add lemon juice and vanilla. Pour into graham cracker pie crust. Chill 3 hours to set. Serve topped with cherry, raspberry, or blueberry pie filling. Or, add frozen berries to Danish Dessert for topping.

Fruit Cobbler

½ cup butter

2 cups flour

1 tablespoon baking powder

1 cup sugar

1½ cups milk (or reconstituted powdered milk or equal parts evaporated milk and water)

1 quart bottled fruit, or equal amount canned fruit, such as peaches

Preheat oven to 350° F.

Melt butter in a 9x13-inch pan. Mix dry ingredients together and stir into melted butter. Add milk and whisk until blended. Spread evenly in pan. Pour bottled fruit or canned fruit, including juice, over top of batter. Bake until brown, about 40 minutes.

The following is a list of some common measurement conversions.

SIMPLIFIED MEASURES

Dash = less than ⅛ teaspoon

3 teaspoons = 1 tablespoon

1 cup = ½ pint

2 cups = 1 pint

2 pints = 4 cups = 1 quart

4 quarts = 1 gallon

1 cup = 8 fluid ounces = 16 tablespoons

¾ cup = 6 fluid ounces = 12 tablespoons

⅔ cup = 5⅓ fluid ounces = 10⅔ tablespoons

½ cup = 4 fluid ounces = 8 tablespoons

⅓ cup = 2⅔ fluid ounces = 5⅓ tablespoons

¼ cup = 2 fluid ounces = 4 tablespoons

⅛ cup = 1 fluid ounce = 2 tablespoons

WORKSHEETS

Our worksheets take the guesswork out of food storage and offer you a step-by-step method that makes it easy for you to get started on your three-month supply. Worksheets are included so that you can plan your meals, make your shopping list for each meal, and then log your purchases by quantity and date. Full-sized worksheets can be downloaded free from our Web site, http://www.notyourmothersfoodstorage .com. Store worksheets in a binder for easy reference.

Meal Planner Worksheets—There are extra worksheets for breakfast, lunch, and dinner meals. Use these worksheets to help you plan the meals your family will eat during a three-month period. You can also use the worksheets to plan for longer periods of time, such as six months, or one year. You should have already practiced using the meal planning worksheets as you were reading this workbook, so you know how easy they are to use.

Shopping List Worksheet—The shopping list worksheet will help you know exactly what you need to buy for your food storage. We suggest using a separate worksheet for each category, such as meals for breakfast, lunch, and dinner, staples and condiments, and nonfood items, to make it easier to find specific items when you are recording your purchases. When you find an item on sale and can put it together with coupons, buy as much as you can and write down the quantity and date. When you use something from your food storage, put a check mark beside the item to signal you to watch for that product to go on sale so that you can replace what you have used.

BREAKFAST MEAL PLANNER

Breakfasts	# times per month	× 3 months	× number in family	= total number of servings

BREAKFAST MASTER SHOPPING LIST

Write your food storage items in the "Item" column. Write the amount you need for a three-month supply in the "Need to Purchase" column. Each time you purchase the item, enter the quantity and the date purchased.

NEED TO PURCHASE	ITEM	QTY	DATE	QTY	DATE

LUNCH MEAL PLANNER

Lunches	# times per month	× 3 months	× number in family	= total number of servings

LUNCH MASTER SHOPPING LIST

Write your food storage items in the "Item" column. Write the amount you need for a three-month supply in the "Need to Purchase" column. Each time you purchase the item, enter the quantity and the date purchased.

NEED TO PURCHASE	ITEM	QTY	DATE	QTY	DATE

DINNER MEAL PLANNER

Dinners	# times per month	× 3 months = number of times meal is served

DINNER MASTER SHOPPING LIST

Write your food storage items in the "Item" column. Write the amount you need for a three-month supply in the "Need to Purchase" column. Each time you purchase the item, enter the quantity and the date purchased.

NEED TO PURCHASE	ITEM	QTY	DATE	QTY	DATE

DESSERT AND TREAT PLANNER

Desserts and Treats	# times per month	× 3 months = number of times dessert is served

DESSERT & TREAT MASTER SHOPPING LIST

Write your food storage items in the "Item" column. Write the amount you need for a three-month supply in the "Need to Purchase" column. Each time you purchase the item, enter the quantity and the date purchased.

NEED TO PURCHASE	ITEM	QTY	DATE	QTY	DATE

STAPLES AND CONDIMENTS WORKSHEET

Staples and Condiments	How much we use in 1 week	× 4 = 1-month supply	× 3 = 3-month supply	Quantity	Date

NONFOOD ESSENTIALS WORKSHEET

Nonfood Essentials	How much we use in 1 week	× 4 = 1-month supply	× 3 = 3-month supply	Quantity	Date

SUBJECT INDEX

RECIPE INDEX

Recipe Index

Recipe Index

ABOUT THE AUTHORS

Kathy Bray is a retired businesswoman. She currently works as a part-time church service missionary at the South Area Family History Training Center in Orem, Utah. She and her late husband, Raymond, lived for many years in northern California before moving to Utah in 1986. She is the mother of six children, grandmother of twenty-three, and great-grandmother of ten.

Jan Barker grew up in northern California before moving to Utah to attend BYU. She and her husband, Jeff, live on seventy acres outside of Payson, Utah, and have six children and ten grandchildren. Jan is a businesswoman and an artist. She also has twenty-four years of volunteer service in the Boy Scouts of America where she learned to "Be Prepared." She owns American Beauty Academy in Payson with her husband and one of her daughters.